Sebastian Seiffert, Julian Thiele
Microfluidics

Also of Interest

Optofluidics.
Process Analytical Technology
Rabus, Sada, Rebner, 2018
ISBN 978-3-11-054614-9
e-ISBN 978-3-11-054615-6

Emulsions.
Formation, Stability, Industrial Applications
Tadros, 2016
ISBN 978-3-11-045217-4
e-ISBN 978-3-11-045224-2

Nanostructured Materials.
Applications, Synthesis and In-Situ Characterization
Terraschke (Ed.), 2019
ISBN 978-3-11-045829-9
e-ISBN 978-3-11-045909-8

Biomimetic Nanotechnology.
Senses and Movement
Mueller, 2017
ISBN 978-3-11-037914-3
e-ISBN 978-3-11-037916-7

Polymeric Surfactants.
Dispersion Stability and Industrial Applications
Tadros, 2016
ISBN 978-3-11-048722-0
e-ISBN 978-3-11-048728-2

Sebastian Seiffert, Julian Thiele

Microfluidics

Theory and Practice for Beginners

DE GRUYTER

Authors
Prof. Dr. Sebastian Seiffert
Johannes Gutenberg University Mainz
Institute of Physical Chemistry
Duesbergweg 10–14
D-55128 Mainz
Germany
sebastian.seiffert@uni-mainz.de

Dr. Julian Thiele
Leibniz Institute of Polymer Research Dresden
Institute of Physical Chemistry & Polymer Physics
Hohe Str. 6
D-01069 Dresden
Germany
thiele@ipfdd.de

ISBN 978-3-11-048777-0
e-ISBN (PDF) 978-3-11-048770-1
e-ISBN (EPUB) 978-3-11-048784-8

Library of Congress Control Number: 2019934970

Bibliographic information published by the Deutsche Nationalbibliothek
The Deutsche Nationalbibliothek lists this publication in the Deutsche Nationalbibliografie;
detailed bibliographic data are available on the Internet at http://dnb.dnb.de.

© 2020 Walter de Gruyter GmbH, Berlin/Boston
Cover image: Sebastian Seiffert and Julian Thiele
Typesetting: Integra Software Services Pvt. Ltd.
Printing and binding: CPI books GmbH, Leck

www.degruyter.com

For Levke, Lea, and Arved

Preface

Microfluidics is the fine art of creation and manipulation of small portions of fluids, often realized by flow within small, sub-millimeter-scale channels. These small dimensions allow the fluid flow to be controlled with exquisite precision, thereby paving a path for use of microfluidics as a fundament for a multitude of applications in the analytical and preparative sciences, covering a whole alphabet of fields: analytics, biophysics, chemical engineering, diagnostics, and so on. Further use of this method in the realm beyond academic research, such as that of commercial fine-chemical production, sensor development, or encapsulation of active substances, is presently being explored. Microfluidics, though, is a demanding field and working area that requires both profound theoretical understanding and experimental skills. As there is no standard in education and training on this method in many fields of study that actually can benefit most from this technique, we reasoned that an introductory textbook might be helpful. This is what the present book is about. It addresses late-stage students or early-stage researchers, typically at the transition from university studies to PhD research, in fields such as biotechnology, cell biology, chemistry, or chemical engineering. Its prime target is to serve beginners, people who may have read papers or listened to talks highlighting the benefits of microfluidics and now desire (or have been told by their advisors) to apply this method to their own research. The present textbook is targeted at helping these people by introducing the basic physics and working principles of the method, along with introducing its practical implementation. We, the authors, both trained as chemists and specialized in the discipline of polymer science, have gained most of the knowledge laid down in this book by self-experience and by training from colleagues during the times of our PhD and postdoctoral research at Harvard University in the group of David A. Weitz (further refined by additional research with Wilhelm T. S. Huck at Radboud University Nijmegen for J. Thiele). We would like to express our gratitude to our main teachers on microfluidics during these times, who were Adam R. Abate, Mark B. Romanowsky, and Thomas Pfohl (formerly at MPI for Dynamics and Self Organization).

Mainz and Dresden, Fall 2019

Aim and approach of this book

Microfluidics has proven to be of versatile utility in both academic and industrial laboratories, with still further promise. It is therefore worthwhile to consider exploration of its rich potential in a plethora of research facilities. As positive as this sounds, though, it confronts scientists and engineers with a challenge: how to get acquainted with this technique? This is where this book comes into play.

It is the aim of this textbook to give beginners a concise and convincing understanding of the physical-chemical principles of microfluidics, along with further giving a handy and helpful practical guideline to practice one's first (and second) own experiments. In Chapter 1, we detail about the fundamental physics of fluid transport, along with the basics of further areas that are relevant for microfluidics. Chapter 2 then gives an introduction to a prime branch of microchannel-based fluidics, which is that of segmented fluid streams with immiscible components (also named droplet-based microfluidics). Finally, Chapter 3 gets to the heart of this book's intent and presents practical guidelines to the assembly and operation of microfluidic devices, both for droplet-based microfluidics and for the flow of continuous fluid streams with miscible components (also named co-flow microfluidics). A special emphasis is on troubleshooting and acquiring competence for conceptual understanding about which experimental conditions will lead to which type of flow in a microchannel system.

The target group of all the matter introduced in the three main chapters of this book is diverse. As readers from different disciplines or with different intent of reading this book may have different interest in following detailed derivations of equations, in Chapter 1, each equation that is a central cornerstone of practical utility is framed, whereas those "only" serving as the way to get to these cornerstones are given without such special emphasis. With this approach, we ensure that each equation is well comprehensible and does not just appear from nowhere, whereas readers with a more pragmatic intention may skip these parts and just consider the framed mathematical milestones.

In the Preface section, it is stated that "there is no standard in education and training on this method [meaning: in microfluidics]". Yet, there is fantastic standards in the education and training on the fundamental physics that lays the fundament for this method. Part of that standard is outlined in Chapter 1 of this book, partially following and inspired by the didactical lines of the textbooks "Experimentalphysik 1" by W. Demtröder (Springer, 1998), "Physikalische Chemie" by P. W. Atkins (Wiley, 1996), and "Polymer Physics" by M. Rubinstein and R. H. Colby (Oxford University Press, 2003), which are the authors' sources of knowledge on elementary fluid dynamics, molecular motion, and polymer viscoelasticity.

Contents

Preface —— VII

Aim and approach of this book —— IX

Abbreviations —— XV

Volume conversion —— XVII

1 Theoretical background —— 1
1.1 Fluid physics on the microscale —— 1
1.1.1 Ideal fluids —— 1
1.1.2 Viscous fluids —— 5
1.1.3 The physical fundament of microfluidics: Navier–Stokes equations and Reynolds number —— 10
1.1.4 Computational fluid dynamics simulation —— 14
1.2 Complex fluids —— 22
1.2.1 Viscous flow: Microscopic view and dimensionless numbers —— 22
1.2.2 Newtonian and non-Newtonian flow —— 27
1.2.3 Shear thinning —— 28
1.2.4 Polymer solutions —— 29
1.2.5 Colloidal dispersions —— 34
1.2.6 Shear thickening —— 35
1.2.7 Even more complex fluids —— 37
1.3 Diffusion —— 37
1.3.1 Brownian motion —— 37
1.3.2 Fick's first law —— 38
1.3.3 Einstein equation —— 41
1.3.4 Fick's second law —— 43
1.3.5 Random walks —— 45
1.4 Interfacial and capillary forces —— 51
1.4.1 Interfacial tension —— 51
1.4.2 Cohesive versus adhesive forces: Wetting and capillarity —— 56
1.4.3 Interfacially active substances: Surfactants —— 58
1.5 Microfluidic flow patterns —— 63
1.5.1 Co-flow and segmented flow —— 63
1.5.2 Segmented flow: Viscous versus interfacial drag —— 64
1.5.3 Dripping and jetting —— 67
1.5.4 Dynamics of droplet interior and interfaces —— 70
References —— 72

2	**Segmented-flow or droplet-based microfluidics — 77**
2.1	Basics — 77
2.2	Drop-by-drop formation in microfluidics — 77
2.2.1	Single emulsions — 77
2.2.2	Mixing phenomena in microdroplets — 82
2.2.3	Higher-order emulsions — 84
2.2.4	Modes of droplet formation: One-step emulsification versus stepwise emulsification — 88
2.3	Experimental platforms for droplet formation — 96
2.3.1	Glass microcapillary assemblies — 96
2.3.2	Poly(dimethylsiloxane) channel networks — 97
2.3.3	Dominance of wetting in PDMS devices — 98
2.4	Application of segmented-flow microfluidics — 102
2.4.1	Droplet-based microfluidics in analytics — 102
2.4.2	Droplet-based microfluidics for particle design — 104
2.4.3	Droplet-based microfluidics for vesicle design — 110
2.4.4	Droplet sizes and limitations — 118
2.5	Numbering up — 122
	References — 131
3	**Practical realization of microfluidics — 139**
3.1	Fundamentals of microfluidic device design and fabrication — 140
3.1.1	Materials selection guide for microflow cell fabrication — 140
3.1.2	PDMS-based microfluidics: Devices by computer-aided design — 142
3.1.3	PDMS-based microfluidics: Basic channel-system elements — 149
3.1.4	PDMS-Based microfluidics: Preparation of microfluidic master devices by photolithography — 158
3.1.5	PDMS-based microfluidics: Device replication by soft lithography — 163
3.1.6	PDMS-based microfluidics: Assembling microfluidic devices from microchannel replicas and substrates — 166
3.1.7	Glass-capillary microfluidics: Manual fabrication of flow cells — 173
3.2	Running microfluidic experiments — 181
3.2.1	Sample preparation for microfluidic experiments — 181
3.2.2	Moving fluids through microfluidic devices — 184
3.2.3	To-the-world connection of microfluidic devices — 190
3.2.4	Optical inspection of microfluidic devices: Choose of microscopes and cameras — 193
3.2.5	Starting microfluidic experiments — 197
3.2.6	Running microfluidic experiments — 201
3.2.7	Finishing microfluidic experiments — 202

3.3	The practice of continuous-flow microfluidics —— **204**	
3.3.1	Single-phase CF microfluidics: Analytics, sensing, synthesis —— **204**	
3.3.2	Multiphase CF Microfluidics: Mixing, gradients, synthesis, nucleation, self-assembly, surface patterning —— **208**	
3.4	The practice of segmented-flow microfluidics —— **216**	
3.4.1	Device design and tailoring —— **216**	
3.4.2	Formation of W/O microdroplets and polymer microparticles —— **232**	
3.4.3	Beyond simple emulsion droplets: Double and higher-order emulsions —— **244**	
3.4.4	Beyond simple aqueous droplets: Complex fluids in SF microfluidics —— **245**	
3.4.5	Formation of vesicles from double-emulsions templates —— **247**	
3.5	Trouble shooting in microfluidics —— **250**	
	References —— **270**	

Appendix —— **273**

Index —— **275**

Abbreviations

ATP	Adenosine triphosphate
ATPS	Aqueous two-phase system
BF	Bright-field
BSA	Bovine serum albumin
CAD	Computer-aided design
CF	Continuous flow
CFD	Computational fluid dynamics
CLSM	Confocal laser scanning microscopy
CNC	Computerized numerical control
COC	Cyclic olefin copolymer
CP	Continuous-phase (flow)
DC	Direct current
DI	Deionized (water)
DLS	Dynamic light scattering
DM	Drop maker
ESI	Electron-spray ionization
ETPTA	Trimethylolpropane ethoxylate triacrylate
FE	Finite element
FEM	Finite element method
FITC	Fluorescein isothiocyanate
FRR	Flow-rate ratio
HFE	Hydrofluoroether
HFF	Hydrodynamic flow-focusing
hMSC	Human mesenchymal stem cells
HR	High-resolution
IP	Inner-phase (flow)
LbL	Layer-by-layer
MALDI	Matrix-assisted laser desorption/ionization
MEMS	Microelectromechanical system
MF	Microfluidics
MP	Middle-phase (flow)
MS	Mass spectrometry
MTES	Triethoxy(methyl)silane
MW	Molecular weight
NP(s)	Nanoparticle(s)
O/W/O	Oil-in-water-in-oil
P2VP	Poly(2-vinylpyridine)
PAA	Poly(acrylic acid)
PAH	Poly(allylamine hydrochloride)
PAAm	Polyacrylamide
PBS	Phosphate-buffered saline
PDE	Partial differential equation
PDMS	Poly(dimethyl siloxane)
PEG	Poly(ethylene glycol)
PI	Photoinitiator
PL	Photolithography
PLA	Poly(lactic acid)

https://doi.org/10.1515/9783110487701-203

PNIPAM	Poly(*N*-isopropyl acrylamide)
PSS	Poly(sodium 4-styrenesulfonate)
PTFE	Poly(tetrafluoroethylene)
RGD	Peptide sequence arginine–glycine–aspartic acid (Arg–Gly–Asp)
SAXS	Small-angle X-ray scattering
SEM	Scanning electron microscopy
SF	Segmented flow
SL	Soft lithography
SME	Small and medium enterprises
TEMED	Tetramethylethylenediamine
TEOS	Tetraethyl orthosilicate (IUPAC: tetraethoxysilane)
UV	Ultraviolet
W/O/W	Water-in-oil-in-water
W/W/W	Water-in-water-in-water

Volume conversion

Unit	Abbreviation	In cubic meters (m^3)	
1 femtoliter	fL	0.000000000000000001	1×10^{-18}
1 picoliter	pL	0.000000000000001	1×10^{-15}
1 nanoliter	nL	0.000000000001	1×10^{-12}
1 microliter	µL	0.000000001	1×10^{-9}
1 milliliter	mL	0.000001	1×10^{-6}
1 liter	L	0.001	1×10^{-3}
1 cubic meter	m^3	1	1×10^{0}

1 Theoretical background

1.1 Fluid physics on the microscale

1.1.1 Ideal fluids

Microfluidics is the art and science of fluid flow on small scales. On such scales, features such as efficient heat transfer, diffusion-based mixing, and a large surface-to-volume ratio (which, in turn, governs wetting phenomena as well as fluid resistance) characterize processes, as illustrated in Figure 1.1. The most important feature, however, is that fluid flow in microchannels is purely laminar. In this chapter, we focus on the fundamentals of these principles in microfluidics.

In this context, a first and central question is how such flow occurs, and how it can be modeled and predicted in a quantitative fashion. The usual approach to tackle questions of this kind in physics is to balance over the forces involved or to appraise the sum of energies and then minimize it. We follow the first approach and consider the forces acting on a fluid volume element ΔV with mass $\Delta m = \rho \Delta V$, where ρ is the density of the fluid. There are three major contributors:

- Pressure differences between different locations in the (microfluidic) system, leading to pressure-based forces: $\boldsymbol{F}_p = -\mathbf{grad}\, p\, \Delta V$
- Gravity: $\boldsymbol{F}_g = \Delta m\, \mathbf{g} = \rho\, \mathbf{g}\, \Delta V$
- Frictional forces, $\boldsymbol{F}_\xi$, between layers of the fluid flowing with different velocities $\boldsymbol{v}$

Balancing over these three contributors yields Newton's equation of motion for the mass element $\Delta m = \rho \Delta V$:

$$\boldsymbol{F} = \boldsymbol{F}_p + \boldsymbol{F}_g + \boldsymbol{F}_\xi = \Delta m \boldsymbol{a} = \rho \Delta V \frac{d\boldsymbol{v}}{dt} \qquad (1.1)$$

The forces involved in eq. 1.1 may have different magnitude and direction in each given experimental situation. Depending on that, the fluid flow can be dominated by just one of these contributors, whereas the others may be negligible. For example, if the frictional force $\boldsymbol{F}_\xi$ is small compared to the others, the fluid is named an **ideal fluid**; a typical example is the flow of air relative to a plane wing. By contrast, if $\boldsymbol{F}_\xi$ outweighs the other forces, the fluid is termed **viscous fluid**; typical examples are honey or syrup.

We first consider the case of an ideal fluid and focus on a fluid volume element ΔV that flows with speed $\boldsymbol{v}(\boldsymbol{r},t)$ at a spot $\boldsymbol{r}$ in our system at a point in time t. If there is no change of this velocity over time, the flow is named **stationary**. Even then, however, there might be variation of the flow speed in space, meaning that the volume element considered can have a different velocity once it has moved to a different location in our system. All the vectors $\boldsymbol{v}(\boldsymbol{r},t)$ in all locations $\boldsymbol{r}$ of our system constitute

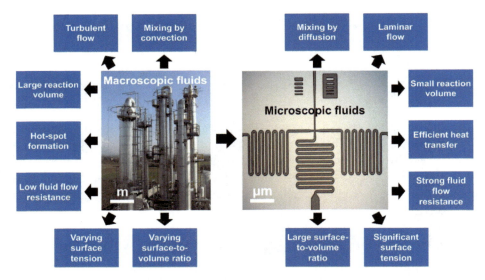

Figure 1.1: Key parameters that come into focus when moving from fluid flow at a macroscopic scale, such as in chemical distillation (left), to fluid flow inside a microfluidic flow cell. Image on the left reproduced from Wikimedia Commons (Luigi Chiesa/CC-BY-3.0).

the **flow field,** which is stationary if all these vectors are time constant. The trajectory *r*(*t*) that our fluid volume element Δ*V* exhibits in the system is named **streamline,** as illustrated for a fluid flowing through a pipe with spatially different diameters in Figure 1.2. In a stationary flow field, the streamlines in the system coincide with the flow field lines, and each trajectory *r*(*t*) follows a corresponding curve **v**(*r*). In microfluidics, the most relevant type of flow is **laminar flow**. In this case, all streamlines are parallel without intermixing; this is the situation sketched in Figure 1.2. The counterpart to laminar flow is **turbulent flow**, which occurs if friction between the

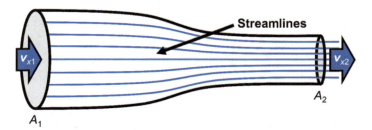

Figure 1.2: Schematic of a circular cross-section microchannel that narrows down from cross-section A_1 at the left end to cross-section A_2 at the right end. A fluid that flows through this channel from the left to the right has a lower flow velocity in the left wide section than in the right narrow section. If the channel diameter is small enough, which it usually is in microfluidics, the flow is laminar, with parallel streamlines.

boundary layers of the fluid and the walls of the channel is strong, whereas friction within the fluid is small compared to the flow-accelerating forces. As to be seen at the end of Section 1.1.2, this type of flow is practically irrelevant in microfluidics, such that we can focus our discussion to the case of laminar flow.

When regarding the experimental situation in Figure 1.2, we can appraise that the mass of a fluid volume element ΔV, which is $\Delta m = \rho \Delta V$, flows through the left-side cross-section A_1 in the form of

$$\frac{dm}{dt} = \rho \frac{dV}{dt} = \rho A_1 \frac{dx}{dt} = \rho A_1 v_{x1} \tag{1.2}$$

If the fluid is incompressible, then ρ is constant, and if there is no way for the fluid to exit the system elsewhere than through the right-side cross-section A_2, we get

$$\rho A_1 v_{x1} = \rho A_2 v_{x2} \leftrightarrow \frac{v_{x1}}{v_{x2}} = \frac{A_2}{A_1} \tag{1.3}$$

The practical meaning of the latter equation is that the fluid flows faster in the narrow right part than in the wide left part of the pipe in Figure 1.2, with an anti-proportional relation between the flow velocities and the channel cross-sections. We can get a more general expression of the latter principle when regarding a fluid volume V of mass

$$m = \rho \int_V dV \tag{1.4}$$

which changes if mass streams out of it or into it. We quantify such a change by appraising the mass streaming across the enwrapping surface of the volume, S, in the direction normal to it, as follows:

$$-\frac{dm}{dt} = \int_S \rho \mathbf{v} d S \tag{1.5}$$

The Gauss theorem allows this surface integral to be converted into a volume integral

$$\int_S \rho \mathbf{v}\, dS = \int_V \mathrm{div}(\rho \mathbf{v}) dV \tag{1.6}$$

thereby yielding

$$-\frac{\partial}{\partial t} \int_V \rho dV = -\int_V \frac{\partial \rho}{\partial t} dV = \int \mathrm{div}(\rho \mathbf{v}) dV \tag{1.7}$$

As this must hold for any volume, we get the **continuity equation**:

$$-\frac{\partial \rho}{\partial t} + \text{div}(\rho \mathbf{v}) = 0 \tag{1.8}$$

which is a generalized mathematical expression of the principle that the mass of a streaming fluid element must be conserved as it flows through a (micro)channel. For incompressible fluids, this further simplifies to

$$\text{div}(\mathbf{v}) = 0 \tag{1.9}$$

Based on all our above recognition, we note that a fluid element may experience acceleration as it flows through a microfluidic channel system. This may either be simply because the fluid volume element gets displaced from one location to another in the system, where different flow-field vectors $\mathbf{v(r)}$ are present, or (potentially on top of that), there may be inherent change of the flow velocity in each location in the system if the flow is nonstationary. In mathematical writing, these two contributors are

$$\frac{d\mathbf{v}_x}{dt} = \frac{\partial \mathbf{v}_x}{\partial t} + \frac{\partial \mathbf{v}_x}{\partial x}\frac{dx}{dt} + \frac{\partial \mathbf{v}_x}{\partial y}\frac{dy}{dt} + \frac{\partial \mathbf{v}_x}{\partial z}\frac{dz}{dt} \tag{1.10a}$$

for the x-component of the flow, with similar expressions for the y component

$$\frac{d\mathbf{v}_y}{dt} = \frac{\partial \mathbf{v}_y}{\partial t} + \frac{\partial \mathbf{v}_y}{\partial x}\frac{dx}{dt} + \frac{\partial \mathbf{v}_y}{\partial y}\frac{dy}{dt} + \frac{\partial \mathbf{v}_y}{\partial z}\frac{dz}{dt} \tag{1.10b}$$

and for the z component

$$\frac{d\mathbf{v}_z}{dt} = \frac{\partial \mathbf{v}_z}{\partial t} + \frac{\partial \mathbf{v}_z}{\partial x}\frac{dx}{dt} + \frac{\partial \mathbf{v}_z}{\partial y}\frac{dy}{dt} + \frac{\partial \mathbf{v}_z}{\partial z}\frac{dz}{dt} \tag{1.10c}$$

which can be expressed together in vector form as

$$\frac{d\mathbf{v}}{dt} = \frac{\partial \mathbf{v}}{\partial t} + (\mathbf{v}\nabla)\mathbf{v} \tag{1.11}$$

with the Nabla operator

$$\nabla \mathbf{v} = \begin{pmatrix} \frac{\partial v_x}{\partial x} & \frac{\partial v_x}{\partial y} & \frac{\partial v_x}{\partial z} \\ \frac{\partial v_y}{\partial x} & \frac{\partial v_y}{\partial y} & \frac{\partial v_y}{\partial z} \\ \frac{\partial v_z}{\partial x} & \frac{\partial v_z}{\partial y} & \frac{\partial v_z}{\partial z} \end{pmatrix} \tag{1.12}$$

With that, the equation of motion of an ideal fluid (with $\mathbf{F}_\xi = 0$) reads

$$\frac{d\mathbf{v}}{dt} = \frac{\partial \mathbf{v}}{\partial t} + (\nabla \mathbf{v})\mathbf{v} = \mathbf{g} - \frac{1}{\rho}\text{grad } p \tag{1.13}$$

which is termed **Euler equation**, as it has been set up by Leonhard Euler back in 1755 already.

1.1.2 Viscous fluids

In the case of a nonideal fluid, frictional forces must be considered as well. We do this by first considering the *origin* of such forces, both from a macroscopic and a microscopic point of view. In a macroscopic perspective, the transport of fluid in x-direction can be modeled and idealized in the form of a drag-induced **Couette flow**. In this picture, a layer of fluid is trapped between two horizontal plates, as sketched in Figure 1.3A. The lower plate is fixed, whereas the upper plate is moved horizontally (that means, in x-direction) at a constant speed $v_x = \partial x/\partial t$. In the limit of slow motion of the top plate, all fluid molecules or particles move parallel to it, and their speed of motion varies linearly from zero at the bottom to v_x at the top. Based on that, the fluid sample can be viewed to be composed of a stack of layers, as also sketched in Figure 1.3A. Each layer moves slower than the one just above it. This is because there is friction between neighboring layers. The stationary lower plate exerts friction to its clinging layer, thereby dragging its motion; this layer exerts consequential friction to the one next above it, thereby dragging its motion in turn, and so forth. You may know a similar experimental situation from your everyday life when pulling a piece of silverware out of a glass of honey. While doing that, some honey will stick to the silverware and be pulled out of the reservoir along with it, exhibiting a parabolic angular profile connecting the flat surface of the honey below and the flat surface of the silverware that stands normal to it. Just as in the idealized picture in Figure 1.3A, this is due to friction that adjacent layers of honey within this profile exert on each other. Based on the same principle, the flow of a fluid within a microchannel, which is the situation of prime relevance in the context of this book, has a parabolic velocity profile, as sketched in Figure 1.3B. In contrast to drag-induced Couette flow, the latter kind of flow is usually pressure-induced, which is referred to as **Poiseuille flow**.

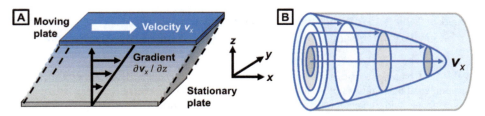

Figure 1.3: (A) Schematic of a Couette flow. A fluid is trapped between two plates, a lower one at rest and an upper one moving with velocity v_x in x-direction. The flow is conceptualized such that the fluid is viewed as a stack of layers, each moving with different velocity, ranging from zero at the bottom layer to v_x at the top layer. The flow velocity gradient that results from this has its physical origin due to the friction between the layers. (B) Based on the same principle, the velocity profile of a fluid pumped through a microchannel such as a cylindrical capillary, which is commonly termed Poiseuille flow, exhibits a parabolic radial profile.

Getting back to the Couette flow sketched in Figure 1.3A: when balancing over the whole system, the magnitude of the frictional force, F_ξ, is proportional to the speed, v_x, and the area, A, of each plate, and it is inversely proportional to the plate–plate separation, z:

$$F_\xi = \eta A \frac{v_x}{z} \tag{1.14}$$

In differential form, the latter equation reads:

$$F_\xi = \eta A \frac{\partial x}{\partial t \partial z} \tag{1.15}$$

The proportionality factor η in eqs. 1.14 and 1.15 is the **dynamic viscosity** of the fluid. Further rearrangement of eq. 1.15 yields a popular variant named **Newton's law**:

$$\sigma = \eta \frac{\partial \gamma}{\partial t} = \eta \dot{\gamma} \tag{1.16}$$

with $\sigma = F/A$ the shear stress, which has the physical dimension of a pressure, and $\gamma = \partial x / \partial z$ the extent of shear deformation.

The dynamic viscosity as introduced in eqs. 1.14–1.16 expresses a fluid's resistance against flow. According to eq. 1.16, its physical S.I. unit is $[\eta]$ = Pa s. Alternatively, a common non-S.I. unit of practical use is dyne s cm^{-2}, also known by its cgs-system name Poise (symbol P) after the French physiologist Jean Poiseuille (1799–1869). Ten poise equals one Pascal second; hence, one centipoise (cP) equals one millipascal second (mPa s). All these units are found as quite a mix in the literature. Fluids that are used in microfluidics typically span a range between about 0.3 mPa s (hexane), 1 mPa s (water), and about 1,400 mPa s (glycerol) at around 20 °C, as compiled in Table 1.1. Note that the viscosities in this table exhibit quite notable decrease upon increase of temperature, as will become understandable below, in Section 1.2.1.

Table 1.1: Dynamic viscosities of some fluids typically used in microfluidics, at three common experimental temperatures. Blue-shaded columns are polar fluids, whereas gray-shaded columns are nonpolar fluids, commonly referred to as "oils" in microfluidics.

T (°C)	η_{Water} (mPas) [1]	$\eta_{Glycerol}$ (mPas) [2]	$\eta_{Chloroform}$ (mPas) [3]	$\eta_{Toluene}$ (mPas) [4]	$\eta_{n\text{-Hexadecane}}$ (mPas) [5]	$\eta_{Squalene}$ (mPas) [6]
20	1.002	1,414	0.573	0.594	3.45	36
25	0.890	905	0.536	0.559	3.06	28
30	0.796	597	0.516	0.527	2.38 [7]	23

> **Note**
>
> A variant of η is the kinematic viscosity:
>
> $$v = \eta \, \rho^{-1}$$
>
> This quantity is a measure of the resistive flow of a fluid under the influence of gravity. Its S.I. unit is $[v] = m^2\,s^{-1}$, but the non-S.I. variant $cm^2\,s^{-1}$ is very common and has been given the cgs-system name Stokes (St) after the Irish mathematician and physicist George G. Stokes (1819–1903). One square meter per second equals ten thousand Stokes.

The internal viscous friction ξ of a fluid that flows through a (microfluidic) channel of radius r at a volumetric flow rate $Q = \partial V/\partial t$ causes a pressure drop Δp along the channel length l, which is proportional to the volume flow with a constant of proportionality named as R_{hydr} according to the **Hagen–Poiseuille law**:

$$\Delta p = \xi \frac{\partial V}{\partial t} = R_{hydr} Q \qquad (1.17)$$

The S.I. units are $[Q] = m^3\,s^{-1}$, $[\xi] = Pa\,s\,m^{-3} = kg\,(m^4\,s)^{-1}$, $[\Delta p] = Pa = kg\,(m\,s^2)^{-1}$. In microfluidics, the Hagen–Poiseuille equation is of practical relevance to calculate the pressure that is necessary to press a fluid through a microchannel with a given cross-sectional shape at a desired flow rate. The proportionality factor R_{hydr} in eq. 1.17 is named hydraulic resistance; Figure 1.4 overviews some typical examples of that quantity.

Note the tremendous dependence on r^{-4}, a^{-4}, or h^{-4} in the channel geometries in Figure 1.4. This can have severe consequences. For example, in view of biology, this drastic dependence causes arteriosclerosis to be a severe condition, as it markedly increases the pressure needed to pump blood through capillary veins. Fortunately, blood's ability to exhibit shear thinning, as detailed further in Section 1.2.7, is highly beneficial in this context, as it allows the blood viscosity to drop in such a constricting environment.

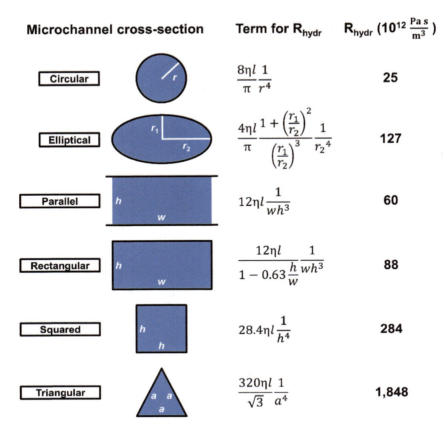

Figure 1.4: Hydraulic resistance R_{hydr} of different kinds of cross-sectional (micro-)channel geometries with channel length l. For illustration, numerical values of R_{hydr} are calculated for water flowing through these geometries, based on the following parameters: r_1 = 50 μm, r_2 = 100 μm, h = 100 μm, w = 200 μm, l = 1 m, η = 1 mPa s. Adapted from H. Bruus, *Lab Chip* 2011, *11*, 3742. DOI: 10.1039/c1lc20658c; copyright 2011 Royal Society of Chemistry.

> **Back of an envelope calculation**
>
> Consider a 1-cm long microchannel with a spherical cross-section of 100 μm diameter, through which water shall be flown at a volumetric flow rate of 1 mL h^{-1} at 25 °C. With the Hagen–Poiseuille equation, the pressure needed for this purpose can be calculated to be:
>
> $$\Delta p = \frac{8 \times 0.89 \text{ mPa s} \times 1 \text{ cm}}{\pi (50 \mu m)^4} \times 1 \text{mL h}^{-1}$$
>
> $$= \frac{8 \times 0.89 \times 10^{-3} \text{Nm}^{-2}\text{s} \times 10^{-2}\text{m}}{\pi (5 \times 10^{-5}\text{m})^4} \times 10^{-6} \text{m}^3\text{h}^{-1} \frac{1}{3{,}600 \text{s h}^{-1}} \approx 1 \text{ kPa}$$
>
> This is easy to realize with devices such as syringe pumps, which are in very common use in practical microfluidics.

How to measure viscosity? (Part 1)

The Hagen–Poiseuille equation gives a quite direct means to experimentally determine the viscosity of a fluid that we may want to use in microfluidics. The principle is to flow a determined volume, ΔV, of the fluid through a tiny capillary of radius r and length l and measure what time, Δt, that takes at a given pressure drop, Δp, along the capillary:

$$\eta = \frac{\pi r^4 \Delta p}{8l} \frac{\Delta t}{\Delta V} = C_{\text{Instrument}} \Delta p \Delta t$$

With $\quad \dfrac{\pi r^4}{8l\Delta V} = C_{\text{Instrument}}$

In the most simple case, no external pressure is applied; the driving force for the fluid flow is then just gravity acting on a vertically aligned capillary, such that the pressure drop along that capillary is simply the hydrostatic pressure of the fluid column:

$$\Delta p = \rho_{\text{Fluid}} |\mathbf{g}| l$$

With that, we may define:

$$C'_{\text{Instrument}} = C_{\text{Instrument}} |\mathbf{g}| l$$

So we get:

$$\eta = C_{\text{Instrument}} \Delta p \Delta t = C'_{\text{Instrument}} \rho_{\text{Fluid}} \Delta t$$

How to measure viscosity? (Part 2)

A common instrument for such experiments is an Ubbelohde viscometer, which consists of a fluid reservoir of defined volume ΔV (typically approx. 5 mL) above a vertically aligned capillary of a determined length (typically approx. 10 cm) and radius, r. A plethora of these instruments is commercially available.

Flowing a defined fluid volume through the capillary allows the viscosity to be determined based on the upper equations. The challenge is the tremendous relevance of the capillary radius in the Hagen–Poiseuille law, which enters with a power of −4. Hence, any imprecision of that radius, and also any inhomogeneity of the capillary cross-section along its length, will cause marked errors when the viscosity is calculated directly based on the upper equation.

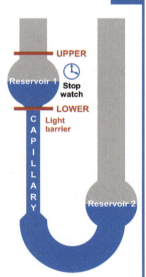

> **How to measure viscosity? (Part 3)**
>
> To circumvent this impairment, a common way to conduct a viscometry experiment is to first estimate a viscometer constant, $C_{Instrument}$ or $C'_{Instrument}$ in the above equations, by using the instrument along with a fluid of known viscosity, such as water, and then to conduct the experiment with the fluid of unknown viscosity.
>
> Even then, a crucial parameter to be controlled as good a possible is temperature, because fluid viscosities usually exhibit quite marked dependencies on that (compare Table 1.1 above). This is commonly achieved by submerging the viscometer in a large water bath of well controlled temperature, ensuring full equilibration of the instrument and sample temperature with that bath by letting it sit in there for at least 30 minutes before conducting the experiment.
>
> In addition to Ubbelohde viscometry, there's a multitude of further methods to measure and quantify viscosities, detailed in further literature such as the Springer Handbook of Experimental Fluid Mechanics or the book Viscosity of Liquids: Theory, Estimation, Experiment, and Data by Viswanath. The advantage of Ubbelohde viscometers is their high precision at comparably low cost.

1.1.3 The physical fundament of microfluidics: Navier–Stokes equations and Reynolds number

Having introduced and clarified to ourselves the role of viscosity, we may now fully appraise the three contributors to eq. 1.1:
- Pressure-based forces: $d\boldsymbol{F}_p = -\mathbf{grad}\, p\, dV$
- Gravitational forces: $d\boldsymbol{F}_g = \rho\, \mathbf{g}\, dV$
- Frictional forces: $d\boldsymbol{F}_\xi = \eta\, \Delta \boldsymbol{v}\, dV$

Along with

$$\frac{d\boldsymbol{v}}{dt} = \frac{\partial \boldsymbol{v}}{\partial t} + (\nabla \boldsymbol{v})\boldsymbol{v} \tag{1.18}$$

we get the **Navier–Stokes equation**:

$$\rho\left(\frac{\partial}{\partial t} + (\nabla \boldsymbol{v})\right)\boldsymbol{v} = -\mathbf{grad}\, p + \rho \mathbf{g} + \eta \Delta \boldsymbol{v} \tag{1.19}$$

For ideal fluids, the condition η = 0 turns the latter equation, which is a puzzling second-order differential equation, into the Euler eq. 1.13, which is an easy first-order differential equation.

The right side of eq. 1.19 compiles the forces acting on our fluid, and the left side summarizes its motion caused by these forces. The first term on the right, $\partial v/\partial t$, reflects the temporal change of v at a fixed position r in a case of nonstationary flow, whereas the second term, $(\nabla v)v$, reflects the spatial variation of v that a volume element experiences if it moves within the flow field. Utilizing the mathematical vector relation

$$(\nabla v)v = 0.5 \text{ grad } v^2 - (v \cdot \mathbf{rot}(v)) \tag{1.20}$$

clarifies that the spatial variation of v can be split into two parts, one representing the change of the absolute of v and the other representing the change of its direction. The latter may lead to formation of vortices in the flow, which become relevant at dominance of inertial forces. The onset of transition between laminar flow, as sketched in Figure 1.5A, and turbulent flow with vortices, as sketched in Figure 1.5B, is given by a **dimensionless number** balancing inertial versus viscous forces: the Reynolds number, $Re = \rho v_x L_0 \eta^{-1}$.

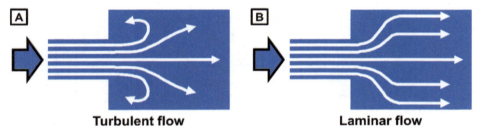

Turbulent flow   **Laminar flow**

Figure 1.5: Streamlines of (A) turbulent flow with vortices and (B) laminar flow with streamlines parallel to the microchannel wall.

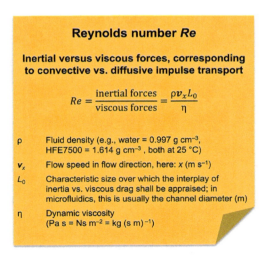

Reynolds number *Re*

Inertial versus viscous forces, corresponding to convective vs. diffusive impulse transport

$$Re = \frac{\text{inertial forces}}{\text{viscous forces}} = \frac{\rho v_x L_0}{\eta}$$

- ρ Fluid density (e.g., water = 0.997 g cm^{-3}, HFE7500 = 1.614 g cm^{-3}, both at 25 °C)
- v_x Flow speed in flow direction, here: x (m s^{-1})
- L_0 Characteristic size over which the interplay of inertia vs. viscous drag shall be appraised; in microfluidics, this is usually the channel diameter (m)
- η Dynamic viscosity (Pa s = Ns m^{-2} = kg (s m)$^{-1}$)

Dimensionless quantities and the Buckingham π theorem (Part 1)

Dimensionless numbers help to predict the interplay of physical quantities and provide information on the physical state of a system – whether fluid flow is laminar or turbulent, for instance. The Buckingham π theorem (BπT) states that we can express any physical relation as identity of a maximal set of independent, dimensionless combinations of variables:

„If there is a process with a certain number of physical variables *n* and *k* independent dimensions, then the original equation describing this phenomenon can be rewritten in terms of a set of dimensionless quantities *p*."

$$p = n - k$$

p: Number of dimensionless quantities (e.g., Re, Ca, We)
n: Number of physical variables (e.g., pressure, velocity, surface tension)
k: Number of physical dimensions (e.g., m, kg, s, K, mol)

Thus, the BπT allows for determining dimensionless quantities from given variables without knowing an exact formula or the legitimacy of a process.

Note: There are, of course, an **infinite number of possibilities** to find independent dimensionless combinations, and the BπT **does not necessarily lead to the physically most meaningful**.

Dimensionless quantities and the Buckingham π theorem (Part 2)

Example of application from technical chemistry and biotechnology:

We consider the challenge of determining the power consumption of a stirred batch reactor, as it is frequently utilized in technical chemistry and cell-based biotechnology (→ fermenter).

Intuitively, power (consumption) P [mass × length² × time⁻³] rises with density ρ [mass × length⁻³] and dynamic viscosity η [mass × length⁻¹ × time⁻¹] of the reactor content to be mixed. Other variables that are relevant include the velocity of stirring ω [time⁻¹] and the size / diameter of the stirrer d [length].

$5 - 3 = 2$	$n = 5 \rightarrow$ power, density, viscosity, speed, size
	$k = 3 \rightarrow$ mass, length, time

With five variables and three dimensions, we need to find two independent dimensionless combinations. **Guided by common practice**, one may choose those as follows:

Reynolds number Re
$$Re = \frac{\rho v_x L_0}{\eta}$$

Newton number Ne
$$Ne = \frac{P}{\rho \omega^3 d^5}$$

A key advantage of dimensionless quantities such as the Reynolds number is that – with just a few experiments based on a defined model – we can utilize these results and describe any other (real-life) experiment, as long as the dimensionless quantity is similar to the one of the model. A key theorem of this so-called similitude concept is the Buckingham π theorem, which is often discussed in the context of dimensionless quantities. This theorem describes that the legitimacy of physical laws does not depend on a unit system.

The *Re* number helps to quantify and delimit different types of fluid flow from one another. *Re* lower than ~250 generally indicates laminar flow. For pipeline-type systems, the range of transition from laminar to turbulent flow is commonly named to be at about *Re* = 2,300. Focusing on water as a typical fluid, and on typical flow velocities of 1 µm s^{-1} to 1 cm s^{-1}, along with typical channel radii of 1–100 µm, the Reynolds numbers in microfluidics range between about 10^{-6} and 10^1. This is markedly smaller than Reynolds numbers of flow processes in our macroscopic world. Thus, in the microscopic world of microfluidics at its inherent low values of *Re*, viscous forces outweigh inertia, and the resulting flows are always laminar. Consequently, the nonlinear terms in the Navier–Stokes equation disappear, leaving the much better solvable **Stokes equation**, which lays a fundament for quantitative prediction and modeling of stationary and laminar flow in microchannels:

$$0 = -\mathbf{grad}\ p + \rho \mathbf{g} + \eta \Delta \mathbf{v} \tag{1.21}$$

A key example for a solution of the Navier–Stokes equation is the steady-state, pressure-driven **Poiseuille flow**, which has been briefly introduced in Section 1.1.2. For a circular microchannel with radius *r* and for a rectangular microchannel with height *h* greater than its width *w*, **v** can be calculated as follows:

Circular channel:

$$v_x(y,z) = \frac{\Delta p}{4\eta l}\left(r^2 - y^2 - z^2\right) \tag{1.22a}$$

Rectangular channel ($h < w$):

$$v_x(y,z) = \frac{4h^2 \Delta p}{\pi^3 \eta l} \sum_{1,3,5\ldots}^{\infty} \frac{1}{n^3}\left[1 - \frac{\cosh\left(n\pi \frac{y}{h}\right)}{\cosh\left(n\pi \frac{w}{2h}\right)}\right] \sin\left(n\pi \frac{z}{h}\right) \tag{1.22b}$$

In eqs. 1.22, *x*, *y*, and *z* correspond to the dimensions of the microchannel along the flow. The characteristic contour lines for the velocity field $v_x(y,z)$ of a Poiseuille flow in a rectangular microchannel is sketched in Figure 1.6.

With the velocity field at hand, we can calculate the volumetric flow rate Q (introduced in Section 1.1.2 in the context of eq. 1.17), which is the fluid volume transported through a (micro-)channel per time interval.

$$Q = \int_{\text{cross-section}} v_x(y,z)\,dy\,dz \qquad (1.23)$$

While the S.I. unit is generally given in $m^3\,s^{-1}$, this is not a very handy unit when it comes to volume flow as found in microfluidic devices, such that in this discipline, the unit $\mu L\,h^{-1}$ is more common. The conversion factor between these units is $1\,m^3\,s^{-1} = 3.6 \times 10^{12}\,\mu L\,h^{-1}$. For circular cross-sectional microchannels, as found in glass-capillary microfluidics (cf. Section 3.1.7), and for rectangular cross-sectional microchannels, as typically found in stamped microfluidics (cf. Sections 3.1.2–3.1.6), the volumetric flow rate can be calculated to be:

Figure 1.6: Lateral view on a velocity field of Poiseuille flow in a rectangular microchannel with $h < w$ (acc. to Figure 1.4). Adapted from H. Bruus, *Lab Chip* 2011, *11*, 3742. **DOI: 10.1039/c1lc20658c**; copyright 2011 Royal Society of Chemistry.

Circular channel:

$$Q = \frac{\pi r^4}{8\eta l}\Delta p \qquad (1.23a)$$

Rectangular channel ($h < w$):

$$Q = \left[1 - 0.63\frac{h}{w}\right]\frac{h^3 w}{12\eta l}\Delta p \qquad (1.23b)$$

Actually, a closed form or analytical solution for a rectangular channel has not been discovered yet. Nevertheless, the approximation for a rectangular channel with $h = w$ (that is, with a *square* cross-section) yields a volumetric flow with an error of just 13%, which further decreases to as little as 0.2% for $h = 0.5\,w$, that is, a microchannel with an aspect ratio of 0.5.

1.1.4 Computational fluid dynamics simulation

The afore-described equations and the dimensionless quantity Re provide a first basis to understand the theory of fluid flow in microfluidic devices. However, for predicting the formation of a specific flow pattern therein, partial differential equations (PDEs) such as the Navier–Stokes eqs. 1.8 and 1.19 need to be solved. This is a challenge, because even for rather simple microflow cells, such as those for miniaturized fluid flow experiments in flow-through analysis, the experimental complexity of the microflow cells has reached a point at which prediction and optimization of experimental conditions is tough. On top of that, although rapid prototyping by combined photo- and soft lithography (all detailed further in Sections 3.1.4–3.1.6) or by additive manufacturing allows for creating microflow cells with a wealth of microchannel geometries and functionalities that can be tested and optimized step by step, loop optimization is hardware- and consumable-intensive. To address this challenge, computational fluid dynamics (CFD) simulations – commonly based on the **finite element method (FEM)** – have become a standard tool for modeling fluid flow; they use numerical methods to solve PDEs that describe the transport of mass, momentum, and energy in moving fluids. The greatest use of numerical modeling is to design *virtual* prototypes of microfluidic devices such to test their performance in a straightforward manner, with the ability to explore the effects of parameters such as the lengthscales, flow rates, and fluid viscosities without actually having to fabricate different devices. For instance, by incorporating the microchannel geometry, flow rates, diffusion coefficients, and reaction kinetics, numerical simulations can provide valuable insight into mixing processes, reactions in flow and at moving as well as static (fluid-wall) interfaces, and aid the efficient design of microflow cells as well as their optimization (cf. Figure 1.7). This way, we assess the performance of microfluidic experiments both theoretically and illustratively and may predict experiments that are otherwise difficult to intuitively predict or to test in reality.

Figure 1.7: Comparison of experimental data and numerical simulation of a continuous-flow microfluidic experiment. (A) An aqueous solution of the fluorescent dye Rhodamine B is flow-focused into a thin jet by dye-free deionized water in a PDMS-based microflow cell. (B) CFD simulation based on the finite element method (FEM) of the experiment in (A). The 2D model is composed of 70,656 finite elements, converging with an error of approx. 10^{-10} after four simulation iterations. (C) Comparison of the simulated velocity field (top) and fluorescence wide field microscopy image (bottom). Due to the symmetry of the experiment, the simulation can be split in half along the flow direction. The scale bars in all panels denotes 50 μm.

To identify necessary ingredients for the numerical solution of PDEs focusing on fluid dynamics, a plethora of commercial software is available, for example, proprietary programs such as ANSYS Fluent and Autodesk CFD. In addition to these, non-commercial software such as OpenFOAM and FeatFlow are available as well as cloud-computing-based software such as SimScale. Without preference for any of those tools, the examples in this book are based on calculations with the commercial software COMSOL Multiphysics®, a module-based platform for FE analysis. The

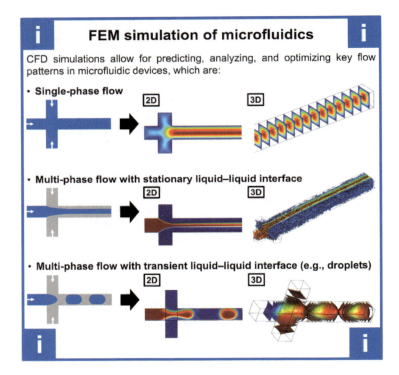

following info boxes focus on key process steps in applying this simulation tool for predicting the flow of incompressible liquids (e.g., water and ethanol) in microfluidic flow-focusing devices.[1] For any system of linear or non-linear differential equations, the typical workflow includes identifying a fluid-dynamics problem ("What do we want to simulate?"), defining equations for it ("How do we want to simulate?"), sketching a flow cell,[2] determining the materials it is made of, and defining the boundaries – e.g., inflow, outflow, and microchannel walls – of the microchannel network ("Where do we want to simulate?"), discretizing the model into FEs, and then choosing suitable solver algorithms for the solution process. Finally, the model can be tested by comparison with experimental data, as illustrated in Figure 1.7. This general workflow is applicable in FEM simulation development independent of the fluid dynamics problem, whether it is a single-phase (e.g., a flow-focusing experiment) or two-phase flow (e.g., droplet formation).

---

1 For simulating non-Newtonian fluid flow, turbulences, as well as non-isothermal flow (e.g., the coupling of heat transfer and fluid flow in a microfluidic channel containing a heating pad), extended sets of equations are required utilizing the so-called k-ε or k-ω turbulence models, for instance.
2 Some simulation softwares include simple design tools that are suited to design microfluidic devices with rectangular and round channel cross-sections. For more sophisticated designs, the experimenter may utilize standard CAD tools such as Autodesk AutoCAD®.

FEM simulation of microfluidics: Navier–Stokes equations (Part 1)

For solving both transient (e.g., droplet formation) and steady-state (e.g., laminar co-flow) models in fluid mechanics, the COMSOL's **Computational Fluid Dynamics** module uses the Navier–Stokes equations:

$$\rho \frac{\partial u}{\partial t} - \nabla \cdot [\eta(\nabla u + (\nabla u)^T)] + \rho(u \cdot \nabla)u + \nabla p = F \quad \text{(Transport equation)}$$

$$\nabla \cdot u = 0 \quad \text{(Continuity equation)}$$

The first equation describes a **velocity field in a fluid by applying Newton's 2nd law of motion to a finite element** (→ FEM). The second equation states that the mass of a liquid flowing into a finite element over a period of time must be balanced by the same mass flowing out.

ρ	Fluid density (kg m^{-3})
u	Flow velocity (m s^{-1}) of a fluid element at position x and time t: $u = u(x, t)$
t	Time (s)
∇	Nabla operator: $\nabla = \left(\frac{\partial}{\partial x}, \frac{\partial}{\partial y}, \frac{\partial}{\partial z}\right)$
η	Dynamic viscosity (Pa s)
p	Pressure (Pa)
F	Large-range force, e.g. gravity (N m^{-3})
T	Transpose of velocity gradient tensor ∇u

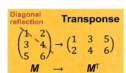

FEM simulation of microfluidics: Navier–Stokes equations (Part 2)

General work flow for solving (fluid dynamics) problems via CFD:

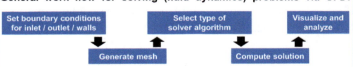

After developing a CAD model of the microflow cell, **boundary conditions are set** to define the interaction of fluid flow with the microchannel walls (→ no-slip boundary conditions). As the **Navier–Stokes equations are computationally challenging**, an appropriate mesh of finite elements in the boundaries of the CAD model is essential. The model is solved for each finite element based on an initial set of variables (e.g., viscosity, flow velocity).

Post processing and visualization:

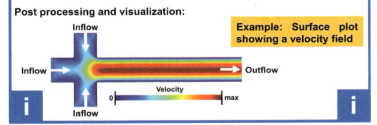

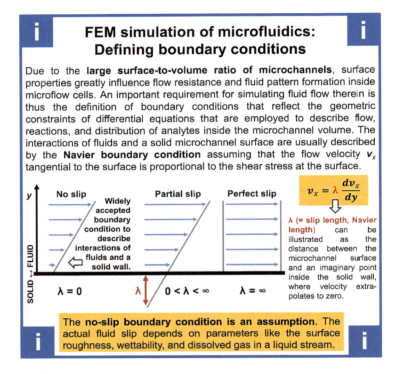

Beyond the slip length on microchannel surfaces (cf. info box above), further boundary conditions need to be defined depending on the function or properties of the microchannel (e.g., being at the inflow or outflow surface) as well as the involved combinations of PDEs. For instance, for a continuous-flow microfluidic experiment forming a continuous jet of one miscible liquid in another one, we require a set of boundary conditions for the Navier–Stokes equations and another set of boundary conditions for mass transport by convection and diffusion. From the many boundaries that need to be defined (cf. Figure 1.8), for the outflow of the microflow cell, a convective flux is a suitable boundary, as convection is proportional to the fluid velocity describing mass transport due to an average velocity of all molecules and bulk fluid motion, respectively.

If the microflow cell is made of soft elastomers and its channels exhibit rectangular cross-sections with a high aspect ratio, these channels are less pressure-resistant than channels with square cross-sections (height = width). As a result, they may expand at high pressure or high flow rates. When studying flow pattern formation and the impact of that channel deformation on the flow profile under these conditions, a linear elastic model (in case of the CFD simulation tool COMSOL Multiphysics®) may be used, and material properties of the flow cell come into focus. If the flow cell is fabricated by PDMS-based soft lithography (cf. Section 3.1.5), we require the

Young's modulus of the PDMS material, E_{PDMS} (approx. 4 MPa), its Poisson ratio, μ_{PDMS} (approx. 0.42), and its density, ρ_{PDMS} (approx. 920 k gm^{-3}).[3]

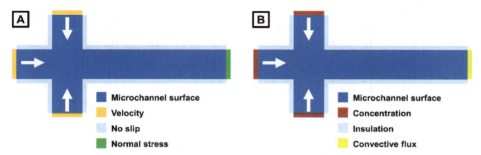

Figure 1.8: Typical boundary conditions in a two-dimensional FEM simulation of diffusion-controlled mixing in a CF flow-focusing experiment (e.g., Figure 1.7). Two sets of PDEs are combined for describing (A) fluid dynamics by the incompressible Navier–Stokes equation, (B) diffusion across the border of the individual fluid streams.

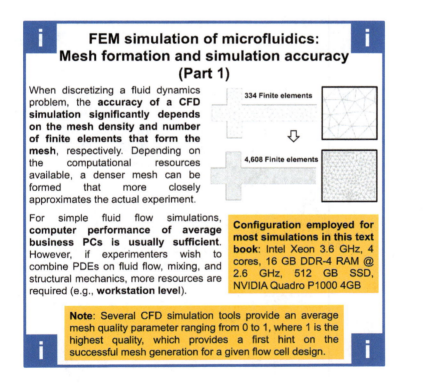

FEM simulation of microfluidics: Mesh formation and simulation accuracy (Part 1)

When discretizing a fluid dynamics problem, the **accuracy of a CFD simulation significantly depends on the mesh density and number of finite elements that form the mesh**, respectively. Depending on the computational resources available, a denser mesh can be formed that more closely approximates the actual experiment.

334 Finite elements

4,608 Finite elements

For simple fluid flow simulations, **computer performance of average business PCs is usually sufficient**. However, if experimenters wish to combine PDEs on fluid flow, mixing, and structural mechanics, more resources are required (e.g., **workstation level**).

Configuration employed for most simulations in this text book: Intel Xeon 3.6 GHz, 4 cores, 16 GB DDR-4 RAM @ 2.6 GHz, 512 GB SSD, NVIDIA Quadro P1000 4GB

Note: Several CFD simulation tools provide an average mesh quality parameter ranging from 0 to 1, where 1 is the highest quality, which provides a first hint on the successful mesh generation for a given flow cell design.

---

3 The values given here are based on PDMS made from a 10:1 mixture of oligomer:crosslinker.

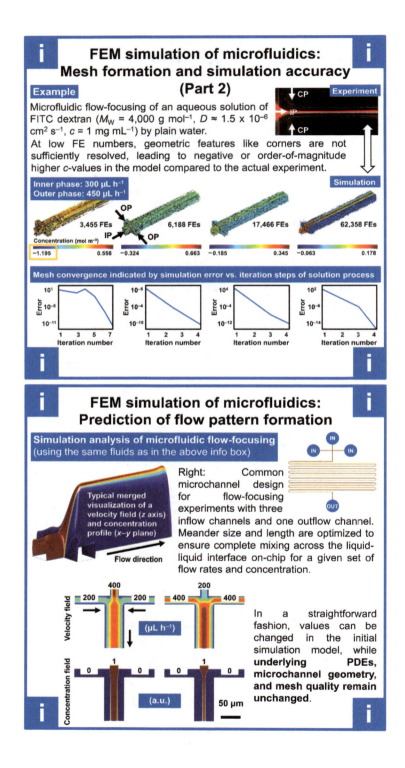

FEM simulation of microfluidics: Optimizing microchannel design (Part 1)

Optimizing diffusion-based mixing in a continuous-flow microfluidics experiment via FEM simulation

Inner phase (FITC dextran, $c = 0.05$ mmol m^{-3}): **0.05 m s^{-1}**
Outer phase (water): **0.03 m s^{-1}**

By tuning the length of the meandering outflow channel, intermixing can be precisely controlled towards complete homogenization upon collection.

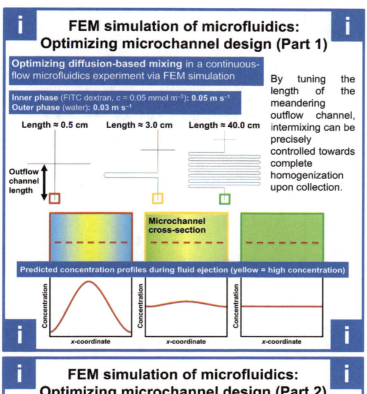

FEM simulation of microfluidics: Optimizing microchannel design (Part 2)

To **prevent adhesion of a dispersed (→ SF microfluidics) or inner phase (→ CF microfluidics) to the surface of a rectangular microchannel** in the absence of a protective sheath flow, its **contact area with the microchannel should be minimized**. Along these lines, CFD simulations support optimization of the microchannel's aspect ratio to minimize the contact area between an inner (e.g., flow-focused) liquid jet and the upper / lower microchannel wall.

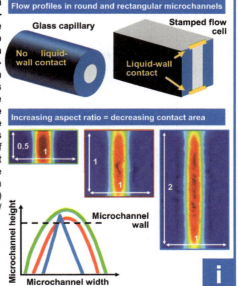

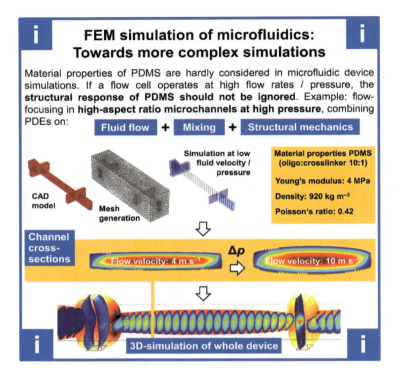

1.2 Complex fluids

Many liquids used in microfluidics-assisted biological, chemical, and technological research, which arcs beyond mere microchannel design and proof-of-principle experimentation, are not as simple as water or oil as listed in Table 1.1. Instead, fluids in the current focus of research include colloidal suspensions, polymer solutions, and biological liquids (e.g., food samples, cytoplasm, cell lysate, blood, or bio-reaction mixtures such as those for PCR and protein synthesis). These fluids are complex, exhibiting properties such as shear thinning, shear thickening, or phase separation. The following section therefore sheds a spotlight on these fluids and discusses – among other parameters – their dissolution, mixing, and flow properties, particularly in view of microfluidic applications.

1.2.1 Viscous flow: Microscopic view and dimensionless numbers

With the continuum perspective on the physics of flow in microfluidic channels, as established in the preceding section, we can model, quantify, and predict the flow of fluids in microchannels. This is sufficiently fine to engineer microfluidic systems. For a deeper conceptual understanding, however, a microscopic view on the role of viscosity is insightful. As a start, we clarify to us that the friction that each layer in Figure 1.3A exerts on its upper-laying neighbor has its physical cause by taking away part of the x-component of the impulse of each molecule or particle in it during each molecular or particulate collision at the layer–layer interface. This is a form of **impulse transport** in the z-direction, which can be captured by a transport equation similar to Fick's first law (that will be treated later in Section 1.3.2.):

$$J_{p,z} = \frac{1}{A}\left(\frac{\partial \boldsymbol{p}_x}{\partial t}\right) = -\eta\left(\frac{\partial \boldsymbol{v}_x}{\partial z}\right) \tag{1.24}$$

If we further follow this thought, we come to the fundamental consideration that flow is nothing else than a form of transport of matter. For such transport, the elementary building blocks of matter – atoms, molecules, or larger entities such as colloidal aggregates, macromolecules, or particles – must be displaced. Such displacement is a process that must overcome an activation barrier. This is because each molecule or particle in a fluid finds itself surrounded by others and must move past these others if it wants to migrate. On top of that, a sufficiently large unoccupied volume element must be close-by for the molecule or particle to migrate into. The interplay of these two effects determines the fluid viscosity, as initially assessed separately by Eyring [8] and Weimann [9], by Doolittle [10], and by Cohen & Turnbull [11], and later combined by Macedo & Litovitz [12]. If no external force is acting on the fluid, the activation barrier has the same height for both the back and forth directions of this elementary process of displacement, as sketched in Figure 1.9A. As a result, the molecules or particles that compose the fluid just randomly displace against each other in the form of Brownian motion, with no preferred direction, as discussed in more detail in Section 1.3. The higher the temperature, the faster is this random motion (as also seen in Section 1.3), because the activation barrier can be overcome more frequently at higher temperature, and because – at isobaric conditions – there is more free volume in the system at higher temperature. If a force is acting in x-direction, however, the molecular or particulate displacement along this direction gets preferred by facing a less high barrier than a displacement in the direction against the external force, as sketched in Figure 1.9B. This is because the directional external force supports the bypassing of neighboring molecules or particles, which, on top of that, now also move into the same direction, thereby further giving way to the molecule or particles next to them. As a result, the molecules or particles that constitute the fluid will preferably displace into the direction of the external force, thereby leading to macroscopic displacement of the entire fluid into that direction.

Appraising the frequencies of the forth- and back displacements in Figure 1.9 can be performed on basis of an Arrhenius-related approach [13], tracing back to original reports by Raman [14] and Andrade [15, 16]:

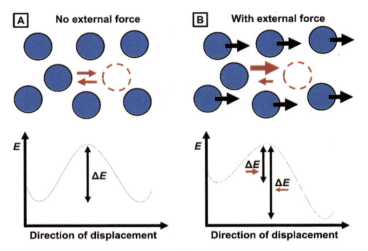

Figure 1.9: Microscopic view on the elementary process of flow. The motion of a molecule or particle in a fluid to a close-by momentarily empty volume element (a "hole") must occur past other close-by molecules or particles, thereby requiring an energy-of-activation barrier, ΔE, to be overcome. (A) In the absence of external forces, this barrier has the same height in the back and forth direction of each elementary molecular or particulate displacement step. As a result, these steps happen with equal likelihood, and hence, with equal frequency, thereby leading to just undirected random motion of the molecules or particles in the fluid. (B) In the presence of an external directing force, the molecular or particulate elementary motion becomes preferred in the direction along the external force, whereas it becomes disfavored in the direction against it. This is because the energies of the formerly equal states change, causing different energies of activation for the molecular or particulate displacement into or against the direction of the external force.

$$\tau^{-1} = f \cdot e^{\frac{-\Delta E}{k_B T}} \tag{1.25}$$

In this equation, τ is a characteristic time that expresses how long it takes for a molecular or particulate displacement to occur. The maximal value for this frequency, f, can be appraised to be similar to the frequencies of vibration of the same molecules or particles in a solid lattice. The energy of activation, ΔE, corresponds to the barriers sketched in Figure 1.9. An upper limit for this quantity can be appraised by the latent heat of vaporization per molecule, as this is the energy required to fully separate the molecules from one another, which must occur against other molecules just like the elementary displacement steps in Figure 1.9. It turns out that ΔE as discussed in our present context is often about a quarter to half of this upper limit [17]. Depending on the extent of the directing external force sketched in Figure 1.9B, and with that, depending on the magnitude of the pressure-driven flow, the transport of matter in microfluidics is dominated by either directed convection or by the ever-present undirected diffusion. The interplay of these two contributors is reflected by another dimensionless number named **Péclet number,** $Pe = v_x L_0 / D$. The role of diffusion in mass transport will be discussed in detail in Section 1.3.

> **Péclet number Pe**
>
> **Convective (flow-based, directed) vs. diffusive (thermal, random) transport of matter**
>
> $$Pe = \frac{\text{convection}}{\text{diffusion}} = \frac{v_x L_0}{D}$$
>
> v_x Flow speed in the flow direction (here: x) (m s^{-1})
> L_0 Characteristic channel size over which the interplay of convection and diffusion shall be appraised; in microfluidic, this is usually the channel diameter (m)
> D Translational diffusion coefficient (m^2 s^{-1})

Note the similarity of the Péclet number to the Reynolds number. Whereas Re quantifies the ratio of convective versus diffusive impulse transport, Pe does so for the ratio of convective versus diffusive transport of the matter itself. When these two numbers are related to each other, in turn, yet another dimensionless number results, the Schmidt number: $Sc = Pe/Re$, which relates the diffusive transport of impulse to the diffusive transport of matter.

> **Schmidt number Sc**
>
> **Diffusive transport of impulse vs. diffusive transport of matter**
>
> $$Sc = \frac{Pe}{Re} = \frac{\eta}{\rho D} = \frac{\nu}{D}$$
>
> η Dynamic viscosity (Pa s = Ns m^{-2} = kg (s m)$^{-1}$)
> ρ Fluid density (e.g., water: 0.997 g cm^{-3}, fluorocarbon oil HFE7500: 1.614 g cm^{-3}, both at 25 °C)
> D Translational diffusion coefficient (m^2 s^{-1})
> ν Kinematic viscosity (m^2 s^{-1})

Depending on the relative importance of the diffusive contributor in the *Pe* number, as quantified by its denominator, the parabolic flow profile shown in Figure 1.3B can be smeared. As a result, an initial thin-striped volume element of interest will first be parabolically distorted and then spread into a channel-filling plug by lateral diffusive smearing. Each thin volume element within that plug will undergo the same process in turn, and as a result, the initially thin-striped volume element of interest will find itself exhibiting a broad smeared Gaussian spatial distribution along the channel. This process is named Taylor dispersion [18–20].

The characteristic time in eq. 1.25, τ, is termed **relaxation time**. It is a measure of how long it takes for the molecules or particles that constitute a fluid to move as far as their own size. It is this elementary microscopic movement that allows macroscopic displacement of the entire fluid medium to set in. In other words, from the point on where each molecule or particle that constitutes a fluid can move at least as far as its own size, the whole fluid can undergo notable macroscopic flow. In many classical fluids, at common experimental conditions, these times are very short, on the order of just 10^{-12}–10^{-10} s [13]. These materials exhibit viscosities in a range as compiled in Table 1.1, with a notable dependence on temperature (as to be understood in view of eq. 1.27 below). In materials composed of polymers or colloids, named **complex fluids** or **soft matter** [21], by contrast, the relaxation timescales are much longer, on the order of milliseconds, seconds, or even minutes to hours. When these materials are observed and probed on timescales shorter than that, they show characteristics of solids, such as exhibiting a measurable initial elastic modulus upon deformation such as shear, G_0. When observed and probed on timescales longer than that, by contrast, they exhibit flow, with a measurable dynamic viscosity, η. This ambivalent character is termed **viscoelasticity**, as it displays both viscous and elastic appearances, just depending on the timescale of the experiment relative to the characteristic relaxation time τ. A fundamental equation relating this all together is

$$\eta = G_0 \tau \tag{1.26}$$

Together with eq. 25, this yields

$$\eta = \frac{G_0}{f} e^{\frac{\Delta E}{k_B T}} \tag{1.27}$$

In view of the latter equation and Figure 1.8, viscoelastic materials have energy-of-activation barriers for molecular or particulate displacement much higher than classical low-viscous fluids. In microfluidics, working with such viscoelastic materials is possible only to a limited extent [22]. Once more, a dimensionless number serves to quantify this, expressing whether it is the intrinsic molecular or particulate relaxation on the timescale τ or the external flow field coming along with the shear rate $\dot{\gamma}$, which corresponds to an external shear-based timescale $\dot{\gamma}^{-1}$, that dominates the dynamics of the soft matter involved. This number is termed **Weissenberg number**, $Wi = \tau\dot{\gamma}$. At low Wi, the molecules or particles have enough time to relax if they are distorted by an external flow with shear rate $\dot{\gamma}$, such that they exhibit their unperturbed shapes or arrangements, whereas at high Wi, such relaxation is impaired by the strong external shear, thereby leading to distorted or disarranged molecules or particles in the fluidic experiment [23].

Weissenberg number Wi

Interplay of internal relaxation of (macro)molecules or particles involved in (micro)fluidic experiments and external shear flow

$$Wi = \frac{\text{elastic forces}}{\text{viscous forces}} = \tau\dot{\gamma}$$

τ Relaxation time (s)
$\dot{\gamma}$ Shear rate (s⁻¹)

1.2.2 Newtonian and non-Newtonian flow

Let us get back to Newton's law according to eq. 1.16. If we plot this in the form of a diagram that features the shear stress, σ, on the ordinate and the shear-deformation rate, $\partial\gamma/\partial t$, on the abscissa, we obtain a graph named **flow curve**. This curve has the dynamic viscosity, η, as its slope. Many fluids exhibit linear flow curves, meaning that their viscosity (= their slope), is constant over the parameter range studied, as illustrated by the middle graph (straight line) in Figure 1.10. These fluids are named **Newtonian** fluids, and their flow is named Newtonian flow. In microfluidics, working with Newtonian fluids is straightforward and simple if their viscosity is not too high. Apart from these simple fluids, many other fluids do *not* display Newtonian flow. In the following sections, we consider two relevant cases: shear thinning and shear thickening.

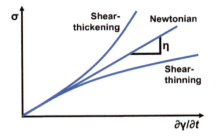

Figure 1.10: Flow curves of a Newtonian fluid (middle straight line) and two non-Newtonian fluids, a shear-thinning (lower curve) and a shear-thickening (upper curve) one.

1.2.3 Shear thinning

In the case of **shear thinning**, the flow curve more and more negatively deviates from the linear form as the shear rate increases, as illustrated by the lower curve in Figure 1.10. With that, the slope of the curve, that is, the fluid viscosity, gets smaller as the shear rate gets stronger. A prime example of such fluids in microfluidics is polymer solutions. These solutions consist of long chain molecules that commonly exhibit shapes of flexible random coils, whereas in special cases, the polymers exhibit shapes of rigid rods. If shear is applied, both these types of objects can get oriented into the direction of flow, as illustrated in Figure 1.11. In the common case of flexible coils, these can also be deformed and elongated in that direction, as also shown in Figure 1.11, provided that the timescale of the external shear, $\dot{\gamma}^{-1}$, is shorter than the time necessary to relax back into a random coil equilibrium structure, τ, which is the case at high Wi. On top of that, if there is coil entanglement in a nondilute solution, shear may also cause disentanglement. All these types of structural change allow the polymers to better displace against one another, and as a result, this change of the fluid microstructure causes a drop in the macroscopic fluid viscosity, a phenomenon that was originally termed **structural viscosity** by Ostwald.

Shear thinning is a phenomenon of expedient practical utility in many applications. For example, wall paint is desired to be low viscous while being brushed to the wall, but after the brushing, it shall be highly viscous to prevent drip. As another example, food like butter is desired to easily spread on your bread, whereas after this application, you desire it to stay in place and not to flow or drip. For purposes such as these (and many others that we could name here), shear-thinning fluids are purposely engineered, and it is often polymers that serve for this purpose. Polymer fluids are also important in microfluidics, because they often serve as precursor fluids to template polymer-based functional microparticles and microcapsules, as discussed in more detail in Chapter 3. They also serve in the fundamental biophysical sciences to mimic natural biological media that are often highly crowded. When working with polymer fluids in microfluidics for these and other purposes, we must be aware of their shear-thinning flow characteristics. In this case, a rule of thumb is that the smaller the microfluidic channel cross-section is, the more severe will shear thinning come into play, as a narrower channel corresponds to greater $\dot{\gamma}$ at a given volume

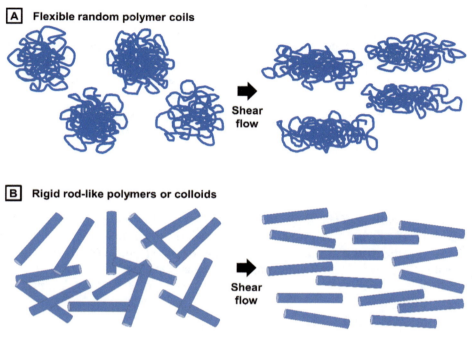

Figure 1.11: Schematic of the structural change in (A) a solution of flexible random polymer coils and (B) a solution of rigid rod-like polymers or colloids. In both cases, the shear gradient orients and also possibly elongates (Panel A) the objects into the direction of flow, thereby allowing them to better slip and slide against one another. The result of this structural change is a decrease of the fluid viscosity upon increase of the shear rate, referred to as shear thinning.

flow rate. This is a significant deviation to the common rule of thumb in microfluidics, after which the flow usually gets more ideal at small channel width if simple Newtonian fluids are involved, because smaller channels cause smaller Reynolds numbers. In polymer fluidics, however, the concurrent gain of importance of shear thinning in smaller channels counteracts this ideality.

1.2.4 Polymer solutions

Even without their complexity of shear-thinning, polymer solutions are tough to handle in microfluidics. This is because they usually have quite high viscosities. To get a deeper impression of that, we consider three relevant types of polymer solutions: first, solutions of polymer chains that are not overlapping, second, solutions of chains that are overlapping but not mechanically entangled, and third, solutions of chains that are both overlapping and entangled, all illustrated in Figure 1.12. These are termed dilute, semidilute unentangled, and semidilute entangled solutions [24], and they are observed one after another as the concentration increases and first passes the threshold for coil overlap, c^* [25], and then the onset for chain entanglement, c_E^*, as also marked in Figure 1.12. Far beyond these two thresholds, another threshold c^{**} exists, whereupon solutions are termed concentrated and exhibit characteristics similar to polymer melts. The overlap concentration c^* can be expressed in different ways [25], a common one being:

$$c^* = \frac{3M}{4\pi N_A R_g^3} \qquad (1.28)$$

with M the polymer molar mass, N_A the Avogadro number, and R_g the radius of gyration of the polymer coils, which is connected to the more common hydrodynamic radius, r_H, that is also indicated in Figure 1.12A, by simple proportionalities depending on the polymer chain architecture and solvency state [26].

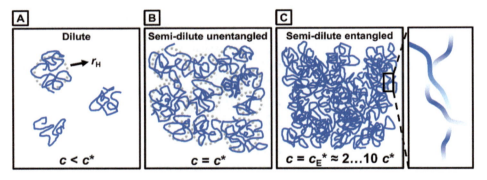

Figure 1.12: Schematic of a polymer solution at (A) dilute concentration, (B) the coil overlap concentration, c^*, upon which the solution is named semidilute unentangled, and (C) the chain entanglement concentration, c_E^*, upon which the solution is termed semidilute entangled. r_H in (A) is the polymer coil's hydrodynamic radius.

Equation 1.26 allows us to estimate the viscosity of a dilute polymer solution. We follow a line of thought by Rubinstein and Colby [24] and appraise the initial elastic shear modulus, G_0, to be in the order of the thermal energy, $k_B T$ (with k_B the Boltzmann constant and T the thermodynamic temperature in Kelvin units) per polymer chain. The number of chains in the solution is estimated as the ratio of its polymer volume fraction, ϕ, and the total segmental volume in each chain, Nl^3, with N the number of segments per chain and l the segmental length. Together, this yields:

$$G_0 \approx k_B T \frac{\phi}{Nl^3} \tag{1.29}$$

With eq. 1.26, we get:

$$\eta_{\text{Dilute}} \approx \eta_{\text{Solvent}} + k_B T \frac{\phi}{Nl^3} \tau \tag{1.30}$$

where τ is the terminal relaxation time of the coil, upon which it exhibits displacement further than its own size; a quantitative expression of this displacement can be obtained by a variant of the Einstein–Smoluchowski equation that will be introduced and discussed in detail below, in Section 1.3.5 (eqs. 1.57 and 1.58): $\tau \approx R^2/D$, were R is the size of the coil and D its diffusion coefficient. The latter quantity can be expressed further by an Einstein equation, as also introduced and discussed in further detail below (Section 1.3.3, eq. 1.46): $D = k_B T / \xi$, where ξ is a friction coefficient. In the Zimm model of polymer dynamics in a dilute solution [27], this quantity is modeled as Stokes-type $\xi = 6 \pi \eta_{\text{Solvent}} R$. Furthermore, a general relation for the coil size is $R = N^\nu l$, with ν the Flory exponent, which is a number between 1/3 and 1 depending on the balance between monomer–monomer and monomer–solvent attractive and repulsive interactions in the solution. Together, the latter expressions give $\tau \approx \eta_{\text{Solvent}} N^{3\nu} l^3 / k_B T$. Plugging this into eq. 1.30 gives the following variant of it:

$$\eta_{\text{Dilute}} \approx \eta_{\text{Solvent}} + \eta_{\text{Solvent}} \phi N^{3\nu-1} = \eta_{\text{Solvent}}\left(1 + \phi N^{3\nu-1}\right) \tag{1.30a}$$

In polymer scientific textbooks, yet another variant of the latter equation is often used:

$$\eta_{\text{Dilute}} \approx \eta_{\text{Solvent}} + \eta_{\text{Solvent}} \phi [\eta] = \eta_{\text{Solvent}}(1 + \phi [\eta]) \tag{1.30b}$$

where $[\eta]$ is the **intrinsic viscosity**, also named **limiting viscosity number** or **Staudinger index**, defined as:

$$[\eta] = \lim_{c \to 0} \frac{1}{c} \frac{\eta - \eta_{\text{Solvent}}}{\eta_{\text{Solvent}}} \tag{1.30c}$$

In this definition, other than in the other equations above, the mass-per-volume concentration of the polymer in solution, c, is used rather than the volume fraction ϕ, but a similar variant of it may also be expressed ϕ, with the only difference that $[\eta]$ in the first case has a physical unit (inverse mass-per-volume concentration), whereas in the second case it has no unit. In the first case, the relation between $[\eta]$ and polymerization degree is often expressed as $[\eta] = K M^a$, which is named Staudinger–Kuhn–Mark–Houwink–Sakurada equation (often shortened to Mark–Houwink equation). Here, M is the polymer molar mass and K and a are empirical constants that are tabulated for many polymer–solvent pairs [28]. Comparison of the Staudinger–Kuhn–Mark–Houwink–Sakurada equation with eq. 1.30a and 1.30b shows that a is related to the Flory exponent, ν, in the form of:

$$a = 3\nu - 1 \tag{1.30d}$$

The range of validity of all the above line of thought is in the regime where the coils have no contact and no mutual interactions to one another, which is given in a rather dilute solution. Upon increase of the polymer concentration, deviations from the above modeling come into play. These may first be captured by a series expansion of eq. 1.30:

$$\eta_{Dilute} = \eta_{Solvent}\left(1 + \phi[\eta] + k_H \phi^2 [\eta] + \ldots\right) \tag{1.30e}$$

with k_H the Huggins constant that accounts for interactions of the coils at concentrations above the highly dilute limit.

Upon further increase of the polymer concentration, we reach a point where the individual polymer coils in the solution get into mutual contact; this happens at the **overlap threshold** or **overlap concentration**, c^* [25]. This concentration scales with the polymer chain length, captured by the number of segments per chain, N, as $c^* \sim N^{-1/2}$ in a θ solvent (coils with ideal size, because coil-expanding and coil-contracting interactions between the polymer and solvent are balanced) and $c^* \sim N^{-4/5}$ in a good solvent (coils expanded by favorable polymer–solvent interactions) [29]. In microfluidics, we often use polymer solutions above c^*, because only then can the polymers be crosslinked with one another such to form a space-filling network, which is often used in conjunction with microfluidic droplet templating such to forming functional size- and shape-controlled polymer microparticles, capsules, and microgels [30]. Furthermore, polymer solutions above c^* can best serve for their other popular application in biophysical microfluidics, where they are used to mimic and study crowding effects like in natural biological fluids [31].

At concentrations above c^*, the polymer coils in a solution interpenetrate and overlap with one another. In a range not too far above this threshold, such interpenetration and overlap does not yet come along with mechanical entanglement. In this range, the polymer-solution viscosity has been reported to be [24]:

$$\eta_{Semidilute-Unentangled} \approx \eta_{Solvent} + \eta_{Solvent} N \phi^{\frac{5}{4}} \quad \text{in a good solvent} \tag{1.31a}$$

$$\eta_{Semidilute-Unentangled} \approx \eta_{Solvent} + \eta_{Solvent} N \phi^2 \quad \text{in a } \theta \text{ solvent} \tag{1.31b}$$

At concentrations of about (2–10) × c^*, mechanical chain entanglement sets in and eventually dominates beyond a threshold c_E^*. In this regime, chain motion and relaxation occurs by a process named reptation [29, 32], which conceptualizes each chain to be trapped in a constraining local environment composed of close-by other chain segments, illustratively viewed as a rigid tube, out of which it can move only by curvilinear motion along its own contour. In this regime, the viscosity of the semidilute entangled solution exhibits severe dependencies on the polymer volume fraction, ϕ, and degree of polymerization, N, both much greater than in eq. 1.31 [33]:

$$\eta_{\text{Semidilute-Entangled}} \sim N^{3.4} \phi^{3.4\ldots3.7} \quad (1.32)$$

From the above set of formulae, we see that the viscosity of polymer solutions exhibits more and more pronounced dependencies on the relevant parameters ϕ and N as we move from the dilute to the semidilute regimes, thereby exhibiting remarkably high values of η. As a result, it gets increasingly tough to work with polymer solutions if their polymer concentration and molar mass increases [22]. On top of that, the phenomenon of shear thinning makes straightforward work with polymer fluids in microfluidics even more complicated, as the effective viscosity is subject to the momentary shear rate applied.

As a practical guideline, we recommend to choose the polymer molar mass and concentration such to be in accord with the goal of the given experiment (for example, such to allow sufficient chain overlap if crosslinking shall be performed in microfluidic droplets to turn them into polymer microgels), but if possible, not higher. That way, the polymer-fluid viscosity can be reduced to the lowest possible yet necessary value according to the preceding discussion. If the resulting viscosity is still too high to achieve controlled microfluidic flow conditions, which is particularly relevant in the variant of droplet-based microfluidics in which monodisperse and size-controlled polymer-fluid droplets shall be formed, there is some tricks that can be applied. One is the use of low-viscous chaperone fluids [34]; another one is the use of polymer-flow-breaking air bubbles in the channel [35]. On top of the complexity of their high viscosities, polymer fluids come along with the complexity of shear thinning, which is particularly relevant in view of the high shear rates that are present in small microfluidic channels. To address this effect, we recommend to choose a large channel size if the given experimental goal permits it. This is because use of larger channels corresponds to lower shear rates at a given volumetric flow rate, thereby minimizing the extent of shear thinning, which, in turn, allows the zero-shear viscosity, η_0, to be used for planning and calculating the characteristic fluid physical parameters of the experiment.

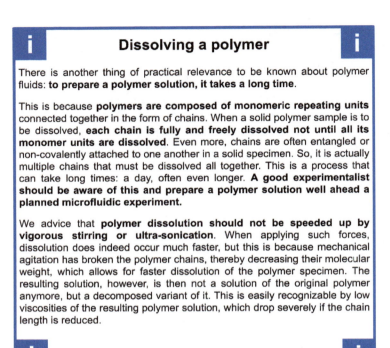

1.2.5 Colloidal dispersions

Another type of complex fluid that has high practical relevance in microfluidics is colloidal suspensions such as nanoparticle dispersions and pretty much all biological fluids. Depending on the specific sort of colloid at hand, the physics of such suspensions can be extremely complex and interesting; and so is their flow. We consider two types of colloids that are both common and still comparably simple. First, we focus on the most simple type imaginable: hard spheres. Second, we focus on colloids of great practical relevance: soft spheres that interact with one another by both van der Waals forces and electrostatics.

Suspensions of **colloidal hard spheres** form a liquid phase at low particle volume fractions. In this limit, the suspension viscosity is quantified by the **Einstein equation**: $\eta = \eta_{Solvent}(1 + 2.5\phi)$ [36, 37], which is nothing else than a variant of eq. 1.30b, with an inherent viscosity of $[\eta] = 2.5$. For colloidal geometries other than spheres, for example ellipsoids, rods, dumbbells, or platelets, a very similar equation expresses the suspension viscosity in the dilute limit, whereby $[\eta]$ reflects the shape of the colloids; for spheres, we have $[\eta] = 2.5$, whereas this number is larger for most other shapes.

Even for the simplest hard-sphere colloids, at high volume fractions, typically $\phi >$ 0.5–0.6, the particles severely block one another, such that their relaxation time, τ, becomes large compared to the experimental timescale. In this limit, the system becomes unable to relax and finds itself in a jammed state named colloidal glass. Already when approaching this jamming volume fraction, microscopic density fluctuations with local glassy dynamics occur in the suspension. Transient trapping of the hard-sphere colloids in these transient local glassy domains manifests itself in the form of a prolonged effective relaxation time, τ_{Eff}. The flow properties of such a non-dilute hard-sphere suspension under shear are governed by that time: if the shear rate is smaller than the inverse of it, local density fluctuations can disappear before they can be perturbed, such that Newtonian flow is observed, as illustrated in Figure 1.13A. If the shear rate and the inverse effective relaxation time are of similar magnitude, however, the local transient glassy domains persist and are deformed such that they are compressed along the compression axis and stretched along the elongation axis of the shear. As a result, flow lines can form in the system, as sketched in an emphasized fashion in Figure 1.13B, thereby facilitating the flow of the suspension in the form of shear thinning.

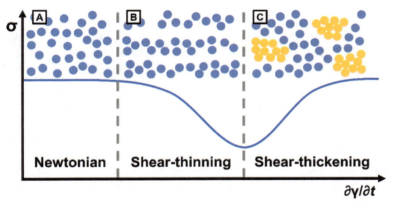

Figure 1.13: Change of the microstructure of a colloidal hard-sphere suspension and its consequential shear thinning and shear thickening. (A) At low shear rates, random collisions between the particles cause Newtonian resistance to flow. (B) As the shear rate increases, transient local glassy clusters can no longer break apart but become deformed in the direction of the shear, thereby creating streamline zones that lower the viscosity. (C) At yet higher shear rates, hydrodynamic interactions between particles dominate over their Brownian motion, thereby spawning hydroclusters (red structures in the sketch). The difficulty of particles to flow around these clusters leads to a notable increase of the viscosity. Adapted from [39]; copyright 2009 American Institute of Physics.

1.2.6 Shear thickening

Within particle-rich compressed domains as just discussed, colloids must overcome mutual hydrodynamic interactions to move away from one another, which manifests itself in the form of a separation time τ_{Sep}. At shear rates smaller than the inverse of this time, the colloids have enough time to detach from one another. At shear rates higher than that, by contrast, the particles form persistent hydroclusters, as illustrated in Figure 1.13C. These clusters markedly impair the flow of the suspension, thereby causing **shear thickening** [38]. The onset of such shear thinning can be estimated by the relation between the particles' Brownian diffusive and their mutual shear-induced hydrodynamic forces, which is again the Péclet number, now given in the variant $Pe = r_H^2 \dot{\gamma}/D$, with r_H and D the hydrodynamic radius and the translational diffusion coefficient of the particles, as both discussed in greater detail in Section 1.3. At low Pe, Brownian diffusion can restore the equilibrium microstructure on the timescale of the shear deformation. At high shear rates, by contrast, deformation of the colloidal microstructure by the flow occurs faster than Brownian motion can restore it. In practice, shear thinning becomes relevant at about $Pe = 1$, whereas at around $Pe = 100$, shear thickening is dominant.

More complex colloids than hard spheres are those that exhibit peculiar interparticle forces. Such colloids often display truly spectacular shear thickening. An illustrative example are mixtures of cornstarch and water. A liquid of a roughly 2:1 volume ratio of these two ingredients can be poured into one's hand, but when squeezed, the liquid morphs into a doughy paste. Upon relieve of the stress, in turn, the pastes reliquefies. Based on this ambivalent nature, there's internet videos showing people running across a large pool of such a mixture, only to sink once they stop in place, and "monsters" that grow out of the mixture when it is acoustically vibrated [39].

Colloids of this kind, and many others that display similar impressive shear thickening, can be viewed as soft-sphere suspensions with electrostatic or steric interparticle repulsion in addition to van-der-Waals-based attraction [40]. Within the framework of the DLVO theory (named after Boris Derjaguin, Lev Landau, Evert Verwey, and Theodoor Overbeek) [41, 42], the sum-potential of these different contributions exhibits two minima, a global one at short interparticle distance and a second local one at longer interparticle distance; in between these two minima lays a maximum that originates from the repulsive contribution. At the absence of external forces, the average interparticle distance is set by the particle concentration. If this concentration is high, particles are forced to be at such close separation that they reach the short-distance global minimum in the sum-potential, thereby causing coagulation of the colloids such to induce a phase-separation into a colloidal crystal phase and a phase that contains just a very few remaining colloids in a dilute suspension. At lower volume fractions, such phase separation may occur in just a local microscopic fashion due to random density fluctuations, and if there is no external force, those locally phase-separating clusters form and break apart again on the experimental timescale, thereby being just transient. If external shear is applied, however, it is again the ratio of the shear rate and the inverse breakup time of such local clusters that determines whether they may become nontransient. This is because under shear, particles are pushed together along the compression axis, just as argued for the hard-sphere case above, and if the shear rate outweighs the inverse breakup time of the resulting clusters, they cannot re-break on the experimental timescale, thereby imparting severe hindrance on flow, which manifests itself in the form of shear thickening [43].

1.2.7 Even more complex fluids

The preceding examples are just two out of a fascinating world of complex-fluid rheology. For instance, seeming variants of shear thinning are so-called Bingham and Casson fluids, which both exhibit a yield point: when subject to external forces such as shear, they first respond like a solid, but at some point, when the force gets strong enough, they flow, either in a Newtonian (Bingham fluid) or in a shear-thinning (Casson fluid) fashion. You may have experienced this phenomenon when spilling ketchup all over your plate upon shaking the bottle. An illustrative picture to explain this phenomenon is that of a house-of-cards structure in the fluid that must be broken before flow can set in. Another variant of both the preceding two cases of shear thinning and shear thickening is fluids that exhibit rheopexy, which is a time-dependent increase or decrease of the viscosity at a constant shear rate.

Many fluids in the biophysical sciences are complex fluids with diverse non-Newtonian characteristics. A prime example is blood, which is both rheopex and shear thinning. One of its major components are discotic cells. These are oriented randomly in the absence of external forces, but they can get oriented parallel to one another at sufficiently strong shear rates, which allows them to better slip and slide against each other, thereby reducing the blood viscosity. This effect is of immense benefit when it comes to blood flow within and through local constrictions [44, 45]. Without detailing further on that, we note that blood and many other biological fluids are, by their very nature, usually both polymer- and colloid-based. As such, they may exhibit a plethora of the above non-simple viscous properties, rendering their use in microfluidics nontrivial and not easily plannable. For a beginner, it will therefore require some time to get to know the specialty of a given complex fluid, including the (limited) range of its (controlled) applicability in microfluidics.

1.3 Diffusion

1.3.1 Brownian motion

Microfluidics is about material transport on small scales. The most elementary mechanism of such transport, however, is diffusion: the random thermal motion of atoms, molecules, and colloids. This ever-present motion will always and unavoidably lead to intermixing of fluids if they have favorable, alike polarities; there is no way to prevent that, not even in the cleverest engineered microfluidic systems. It is therefore the aim of this section to provide a fundament for quantitative assessment of this type of molecular motion, with the goal to quantitatively appraise how relevant diffusive transport is in a given experimental situation. As a practical example, at the end of this section, we aim to appraise how far a protein, which is a type of macromolecular and colloidal matter that is often used in microfluidics, diffuses on a typical experimental timescale, such to appraise the significance of this movement relative to the microfluidic channel dimensions.

Diffusive molecular motion has first been recognized indirectly by the Scottish bonatist Robert Brown, when he observed plant pollen wiggling on the surface of water in 1827 [46]. (Note: that was a time when the concept of atoms and molecules as the basic building blocks of matter was still being established!) Later on, in the 1900s, Albert Einstein and Marian Smoluchowski detailed how that motion results from the steady, permanently unbalanced random impact of surrounding water molecules onto the pollen [47–49].

The mathematics of diffusion is covered within two fundamental laws derived by Adolf Fick (1855) [50], related to similar, earlier established theoretical concepts for the transfer of heat by Joseph Fourier (1822) [51]. Fick's first law establishes a correlation between the (stationary) diffusive flux of a substance to its (stationary) concentration gradient in the system, whereas Fick's second law is a differential equation that relates the spatial concentration gradient to its time-dependent evolution in the system.

1.3.2 Fick's first law

In classical physics, a gradient of a potential, φ, such as the gravitational or electrical potential generally corresponds to a force, $\boldsymbol{F}$, acting down the slope of the potential:

$$\boldsymbol{F} = -\mathbf{grad}\,\varphi \qquad (1.33)$$

For example, the gravitational force corresponds to a gradient in a gravitational field in z-direction:

$$\boldsymbol{F}_{g,z} = -\mathbf{grad}\begin{pmatrix} 0 \\ 0 \\ mgz \end{pmatrix} = m\boldsymbol{g}_z \qquad (1.33a)$$

In chemistry, the most important potential is the chemical potential, μ, denoting the dependence of the Gibb's free energy on the composition of a system under consideration. When there is no gradient in μ, the system is at equilibrium. By contrast, if there is a gradient in μ, then a chemical reaction and/or transport of matter will flatten it. We consider nonreactive systems, where diffusive transport of matter is the sole process of action. For mathematical treatment, it is useful to adopt the general formalism of eq. 1.33 and introduce a conceptual thermodynamic force that corresponds to the gradient in μ:

$$F_{\text{Thermodyn}} = -\text{\textbf{grad }} \mu \tag{1.34}$$

In a single dimension, x, this simplifies to

$$F_{\text{Thermodyn}} = -\left(\frac{\partial \mu}{\partial x}\right) \tag{1.35}$$

Note that **F** is a conceptual but not necessarily a real force pushing molecules down the slope of the chemical potential. What **F** actually reflects is the spontaneous tendency of matter to spread within a medium until uniform distribution is achieved; this is a direct consequence of the second law of thermodynamics. In this view, a statistical consideration of diffusion is insightful. Figure 1.14 shows a system with two kinds of molecules or particles, blue and gray. Panel A displays a nonequilibrium situation, where most of the gray molecules or particles sit in the lower half of the system, whereas most of the blue molecules or particles populate the upper half. As each molecule or particle moves randomly, the likelihood of a gray one to move from the lower to the upper half is higher than that of the opposite case of a gray one moving from the upper to the lower half in Figure 1.14A. This is simply because there are more gray ones in the lower than in the upper half in the momentary situation displayed in this figure, and therefore, even though each individual gray molecule or particle moves by a nondirected and unpredictable random walk, there will be an effective upward overall flux of gray molecules or particles. The same argument holds for the blue ones: it is more likely for a blue molecule or particle to move from the upper to the lower half in the moment depicted in Figure 1.14A than vice versa. Again, this is simply because there are more blue ones in the upper than in the lower half at the moment displayed, and because they all exhibit isotropic random motion with no preferred direction. The overall flux of matter will eventually lead to an equilibrium situation with an equal number of gray and blue molecules or particles in the upper and in the lower half of the system, respectively, as depicted in Figure 1.14B. In that situation, an equal number of gray and blue ones will cross the middle plane from above and below at each time, respectively, again due to their individual equally random motion. The thermodynamic force introduced in eq. 1.35 is a conceptual quantity expressing these statistics. It thereby provides a formalism to quantify the entropy-driven tendency of the system shown in Figure 1.14A to evolve into the equilibrium state shown in Figure 1.14B based on an effective flux of the molecules, which results from nothing else than the uncoordinated random motion of each individual molecule and their initial uneven distribution in the system.

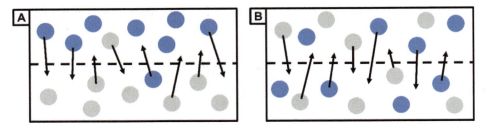

Figure 1.14: Schematic of the random motion of two diffusing species, blue and gray, in a system at (A) nonequilibrium inhomogeneous and (B) equilibrium homogeneous spatial distribution of them. In situation (A), the unequal number of molecules or particles of each kind passing the imaginary middle plane from below or above, respectively, will cause their inhomogeneous spatial arrangement to homogenize with time. In situation (B), the equal number of gray and blue species passing from above and below will no further change their homogeneous spatial arrangement in the system.

The chemical potential of a substance dissolved in a solvent at concentration c is

$$\mu = \mu_0 + RT \ln c \tag{1.36}$$

with μ_0 the standard chemical potential, R the gas constant, and T the thermodynamic temperature. With that, the thermodynamic force acting on the substance is

$$F_{\text{Thermodyn}} = -RT \left(\frac{\partial \ln c}{\partial x} \right) = -\frac{RT}{c} \left(\frac{\partial c}{\partial x} \right) \tag{1.37}$$

or, when expressed in a non-molar form (with the Boltzmann constant k_B):

$$F_{\text{Thermodyn}} = -k_B T \left(\frac{\partial \ln c}{\partial x} \right) = -\frac{k_B T}{c} \left(\frac{\partial c}{\partial x} \right) \tag{1.37a}$$

This force is connected to the experimentally observable flux of the substance considered, J_n, which is quantified by the amount of molecules or particles of that substance transported per time, (dn/dt), through unit area A:

$$J_n = \frac{1}{A} \left(\frac{dn}{dt} \right) \tag{1.38}$$

The molecules or particles reach a constant diffusive drift velocity, v_D, when $F_{\text{Thermodyn}}$ equals the counteracting frictional force, F_ξ:

$$F_{\text{Thermodyn}} = F_\xi = \xi v_D \tag{1.39}$$

In eq. 1.37, ξ is the frictional coefficient, which can be alternatively expressed as the inverse of the molecular or particulate mobility, u:

$$F_{\text{Thermodyn}} = F_\xi = \frac{1}{u} v_D \qquad (1.40)$$

Thus,

$$v_D = u F_{\text{Thermodyn}} \qquad (1.40a)$$

The drift velocity, v_D, is also proportional to the flux, J, as detailed further below in eq. 1.44. Moreover, according to eq. 1.37, $F_{\text{Thermodyn}}$ is proportional to the negative concentration gradient $-(\partial c/\partial x)$. This combination of proportionalities turns eq. 1.40 into

$$J_n \sim -\left(\frac{\partial c}{\partial x}\right) \qquad (1.41)$$

Introducing the diffusion coefficient, D, as a constant of proportionality yields Fick's first law:

$$J_n = -D\left(\frac{\partial c}{\partial x}\right) \qquad (1.42)$$

In a general (coordinate-system independent) form, the latter equation reads

$$J_n = -D\ \mathbf{grad}\ c \qquad (1.42a)$$

The diffusion coefficient has the physical dimension (length)^2 (time)$^{-1}$, that is, $[D] = \text{m}^2\,\text{s}^{-1}$ in S.I. units.

1.3.3 Einstein equation

Our above line of thought has shown that diffusing molecules or particles reach a constant diffusive drift velocity (cf. eq. 1.40a). We now proceed by deriving a relation between v_D and the number of molecules or particles per unit volume:

$$dn = c\,dV = c\,A\,dx = c\,A\,v_D dt \qquad (1.43)$$

Rearrangement of the latter, along with employing eq. 1.38, yields:

$$\frac{1}{A}\left(\frac{dn}{dt}\right) = J_n = c v_D \qquad (1.44)$$

From eq. 1.40a, along with eqs. 1.37a and 1.43, we get:

$$v_D = uF_{\text{Thermodyn}} \Leftrightarrow \frac{1}{c}J_n = u\frac{-k_B T}{c}\left(\frac{\partial c}{\partial x}\right) \Leftrightarrow J_n = u(-k_B T)\left(\frac{\partial c}{\partial x}\right) \quad (1.45)$$

Comparison with eq. 1.42 shows:

$$D = uk_B T = \frac{k_B T}{\xi} \quad (1.46)$$

which is termed **Einstein equation**. This equation shows that the driving force for the diffusive motion of molecules and particles is thermal energy, quantified by $k_B T$ in the numerator on the right side, whereas viscous drag, as quantified by the friction coefficient ξ in the denominator, obstructs this motion. For the limit of spherical particles, the latter coefficient has been detailed to be $\xi = 6\pi\eta r_H$ by George G. Stokes, with η the viscosity of the medium and r_H the hydrodynamic radius of the diffusing objects. This term is often used in eq. 1.46, which is then named **Stokes–Einstein equation**:

$$D = \frac{k_B T}{6\pi\eta r_H} \quad (1.47)$$

This equation allows us to calculate diffusion coefficients if we know the radius of the diffusing molecules or particles and the viscosity of the medium wherein they diffuse. Note, however, that it is not the plain molecule or particle radius that must be used in eq. 1.47, but the *hydrodynamic* radius, which may differ from the actual radius due to contributions like swelling (for example, if the diffusing object is a soft polymer particle) or binding of a hydration shell (especially if the diffusing object is an ion). Vice versa, we can also use eq. 1.47 to estimate the hydrodynamic radius from experimental measurements of the diffusion coefficient. This is what is commonly done in dynamic light scattering (DLS). This experimental method serves to estimate diffusion coefficients and then allows them to be used in conjunction with eq. 1.47 to calculate hydrodynamic radii; this is usually done by the data acquisition and analysis software in situ, thereby implying to beginners that DLS was a technique to directly measure radii. Note that this is misleading. What DLS actually does is to estimate diffusion coefficients and then *calculate* radii from them, assuming that the diffusing objects are spherical and diffuse according to the simple Fickian fundamentals in a plain viscous medium.

1.3.4 Fick's second law

Equation 1.42 holds for situations with a stationary concentration gradient, entailing a corresponding stationary diffusive flux. Nonstationary cases need to be treated by quantifying the diffusive influx and outflux of matter in a volume element, as illustrated in Figure 1.15.

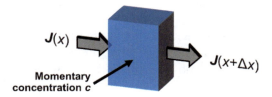

Figure 1.15: Conceptual schematic of the diffusive in- and outflux of matter into and from a volume element with momentary average concentration c.

A time-dependent change of concentration in the volume element results from unequal in- and outflux of mater, quantified as follows:

$$\frac{\partial c}{\partial t} = \lim_{\Delta x \to 0} \frac{-J(x+\Delta x) - J(x)}{\Delta x} = -\frac{\partial J}{\partial x} \qquad (1.48)$$

J can be substituted by eq. 1.42 (Fick's first law), thereby yielding:

$$\frac{\partial c}{\partial t} = -\frac{\partial}{\partial x}\left(-D\left(\frac{\partial c}{\partial x}\right)\right) \qquad (1.49)$$

If D does not exhibit any spatial dependence, then eq. 49 simplifies to:

$$\frac{\partial c}{\partial t} = D\frac{\partial^2 c}{\partial x^2} \qquad (1.49a)$$

Equation 1.49a is **Fick's second law** in a one-dimensional scenario. Its two- and three-dimensional variants are as follows:

$$\frac{\partial c}{\partial t} = D\left(\frac{\partial^2 c}{\partial x^2} + \frac{\partial^2 c}{\partial y^2}\right) \qquad (1.49b)$$

$$\frac{\partial c}{\partial t} = D\left(\frac{\partial^2 c}{\partial x^2} + \frac{\partial^2 c}{\partial y^2} + \frac{\partial^2 c}{\partial z^2}\right) \qquad (1.49c)$$

In all these forms, Fick's second law is a second-order differential equation in space and a first-order differential equation in time. This equation can be solved by accounting for two boundary conditions for the spatial dependence and an additional initial condition for the time dependence, which have to be specified for a given case at hand. Depending on them, various solutions can be obtained, as detailed and compiled in a book by Crank [52]. The following most simple example provides a useful tool for microfluidics in practice. We consider a diffusion process that originates from a sharp delta pulse and then proceeds into an infinite medium, where the total number of molecules is constant at all times.

1. When at time $t = 0$ the substance is localized in an infinitesimally flat but otherwise eternally extended plane in an infinite medium, diffusion occurs in one dimension only (normal to the plane), and the solution to eq. 1.49a is

$$c(r,t) = \frac{M}{2\sqrt{\pi Dt}} e^{\frac{-r^2}{4Dt}} \tag{1.50a}$$

2. When at $t = 0$ the substance forms an infinitesimally thin but eternally long line in an infinite medium, spreading occurs in two dimensions (normal to the line), and one obtains as a solution to eq. 1.49b

$$c(r,t) = \frac{M}{2\pi Dt} e^{\frac{-r^2}{4Dt}} \tag{1.50b}$$

3. When diffusion originates from an infinitesimally small spot source in an infinite three-dimensional medium, the concentration profiles that develop will be spherically symmetric according to the following solution to eq. 1.49c:

$$c(r,t) = \frac{M}{8\sqrt{(\pi Dt)^3}} e^{\frac{-r^2}{4Dt}} \tag{1.50c}$$

In the eqs. 1.50a–1.50c, r represents the generalized (radial) coordinate and M denotes the total amount of the diffusing species in the three-dimensional case, while in the two- or one-dimensional case, it stands for the amount of substance per unit length or unit area, respectively. The three equations can be combined to one by introducing the parameter d for the diffusion dimension:

$$c(r,t) = \frac{M}{\sqrt{(4\pi Dt)^d}} e^{\frac{-r^2}{4Dt}} \tag{1.50d}$$

The concentration profiles given by the latter equations are Gaussians that quantify the spreading of the diffusing substance in a spatially and temporally resolved fashion, as illustrated schematically for a situation that closely matches the idealized one-dimensional case in Figure 1.16. At the beginning, the full amount of the diffusing substance is localized at the origin, which is a plane at the bottom in Figure 1.16. Subsequently, the substance spreads into the medium, which occurs such that just a few molecules or particles quickly displace over long distances, whereas the majority is still localized close to the origin. As time proceeds, this type of spreading spans longer and longer distances. The spatial concentration profile is a Gaussian with $e^{-1/2}$ radius of $(2Dt)^{1/2}$ at each time, with a width that is closely related to another useful quantity: the mean-square displacement, which can be obtained from a statistical treatment, as follows in the next section.

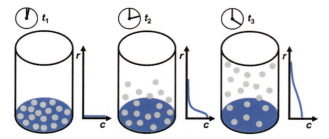

Figure 1.16: One-dimensional diffusive spreading of a substance from a plane source. For simplicity, only the upper half of the system is shown; in the lower half, the same spreading occurs in a mirror-image fashion.

1.3.5 Random walks

An intuitive picture of the mechanism of diffusion is to imagine molecules or particles to move in a series of small steps, thereby gradually migrating away from their original position. This kind of motion can be modeled on a lattice, whereon the molecules or particles undergo sequential random jumps, commonly termed **random walk**. We may imagine such a random walk in one dimension to proceed as if the diffusing molecule or particle would flip a coin before taking each next step, and then go to the right if head or to the left if tail was the outcome of the coin flipping. Analogously, in three dimensions, we can imagine the molecule or particle to roll a dice and then take a step to the left, right, forward, back, up, or down depending on the outcome of the dice rolling. The result is a random sequence of steps. We consider the one-dimensional case for simplicity. In that case, steps can go to the left or right only. We regard a sequence of N such steps in total, N_+ of which to the right and N_- to the left. The end position of this one-dimensional random walk will be $x = N_+ - N_-$. The total number of walks, W, leading to a given end position, x, in N such steps can be appraised by binomial statistics – which also applies for the mathematical treatment of games like coin flipping – as

$$W(N,x) = \frac{(N_+ + N_-)!}{N_+! N_-!} = \frac{N!}{\left(\frac{N+x}{2}\right)! \left(\frac{N-x}{2}\right)!} \quad (1.51)$$

The total number of all imaginable walks is 2^N. Thus, the probability to end up at position x after N steps is:

$$\frac{W(N,x)}{2^N} = \frac{1}{2^N} \frac{N!}{\left(\frac{N+x}{2}\right)! \left(\frac{N-x}{2}\right)!} \quad (1.52)$$

We can use a Gaussian and Stirling's approximation to eliminate the factorials in eq. 1.52 and thereby obtain a much simpler analytical function:

$$\frac{W(N,x)}{2^N} \cong \sqrt{\frac{2}{\pi N}} e^{\frac{-x^2}{2N}} \quad (1.53)$$

If we normalize our coordinate system from a step-width of 1, as we had it so far, to a step-width of ½, we transform the integer distance of the walks' possible end positions on the x-axis from 2 to 1. With that, we get the probability distribution of the displacement x in a one-dimensional random walk:

$$P_{1D}(N,x) = \sqrt{\frac{1}{2\pi N}} e^{\frac{-x^2}{2N}} \quad (1.54)$$

The one-dimensional mean-square displacement is obtained from this distribution by the following operation [24]:

$$\langle x^2 \rangle = \int_{-\infty}^{+\infty} x^2 P_{1D}(N,x) dx = \sqrt{\frac{1}{2\pi N}} \int_{-\infty}^{+\infty} x^2 e^{\left(\frac{-x^2}{2N}\right)} dx = N \quad (1.55)$$

With this result, we can substitute N by $\langle x^2 \rangle$ in eq. 1.54 such to get an equation with just one remaining variable, x:

$$P_{1D}(x) = \sqrt{\frac{1}{2\pi \langle x^2 \rangle}} e^{\frac{-x^2}{2\langle x^2 \rangle}} \quad (1.56)$$

Comparison with eq. 1.50a shows:

$$\langle x^2 \rangle_{1D} = 2Dt \quad (1.57)$$

which is the **Einstein–Smoluchowski equation**. This equation allows us to estimate what average displacement in x-direction diffusing molecules or particles will undergo in a given time, t, or vice versa, how long it will take on average to diffuse a given distance, x. Similar equations hold for the y- and z-directions. In an isotropic situation, where no spatial direction is preferred, the three-dimensional mean-square displacement is obtained by a simple superposition of three individual one-dimensional random walks, one for each direction, as

$$\langle r^2 \rangle_{3D} = 6Dt \qquad (1.58)$$

In the following two boxes, we demonstrate the utility of the Einstein–Smoluchowski equation and appraise the mean displacement that a protein, which is a type of colloidal matter often used in biophysical microfluidics, undergoes in the direction normal to the flow on a typical experimental timescale in a laminar co-flow situation. With that kind of estimation, we may get another view on the Péclet number introduced in Section 1.2.1. To diffuse across the whole channel width, w, the molecules or particles need a time of

$$\tau_D \approx \frac{w^2}{D} \qquad (1.58a)$$

During this time, the directed convective transport down the channel displaces each fluid volume element by a distance x:

$$x = v_x \tau_D = \frac{v_x w^2}{D} \qquad (1.58b)$$

With that, the number of channel widths n over which complete diffusive mixing across the channel width w occurs, is

$$\frac{n}{w} = \frac{v_x w}{D} \qquad (1.58c)$$

This equation is equal to the formerly introduced Péclet number, where the characteristic lengthscale L_0 is just the microchannel width w. In the above example, we have $Pe = 28.571$, so still quite a dominance of convection over diffusion. Nevertheless, our above estimate has shown that even then, quite some remarkable diffusive inter-spreading of the fluid flow occurs. When considering the Stokes–Einstein eq. 1.47, it gets apparent that the relative importance of convective versus diffusive transport further depends on the size of the molecules or particles involved. In a polydisperse sample of differently sized (macro)molecules or particles, all have their individual Pe. Clever adjustment of the channel length then allows their different extent of lateral spreading to be employed for separating them [55, 56].

Back of an envelope calculation

How far does a typical protein move in the direction normal to the flow in a typical microfluidic co-flow experiment? – Part 1

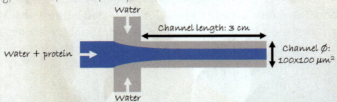

Three aqueous streams flow along each other in a laminar fashion; two outer streams are plain water, the middle stream contains human hemoglobin. In its tetrameric form, this protein has a molar mass of 64,450 g mol^{-1} [53] a radius of about 2.5 nm [54], and a nearly spherical (=globular) shape.

As a first step, we use the Stokes–Einstein equation to estimate the diffusion coefficient of this protein at a temperature of 25 °C (298 K) in water, which has a viscosity of 0.89 mPa s at that temperature (see Table 1.1):

$$D = \frac{k_B T}{6\pi \eta r_H} = \frac{1.38 \times 10^{-23} \text{JK}^{-1} \times 298 \text{K}}{6\pi \times 0.89 \times 10^{-3} \text{Pa s} \times 2.5 \times 10^{-9} \text{m}} = 9.8 \times 10^{-11} \text{m}^2 \text{s}^{-1}$$

$$D = 98 \, \mu\text{m}^2 \text{s}^{-1}$$

For simplicity, in this estimation, we have taken the particle radius as the hydrodynamic one.

Back of an envelope calculation

How far does a typical protein move in the direction normal to the flow in a typical microfluidic co-flow experiment? – Part 2

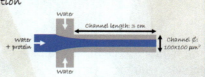

As a second step, we estimate the contact time of the inner and outer streams in the sketch above. We consider a rectangular microchannel that is 3 cm long, with a uniform width and height of 100 μm each. All fluids are pumped with volumetric flow rates of 1/3 mL h^{-1} each, corresponding to a total volumetric throughput of 1 mL h^{-1} = 1 cm^3 h^{-1}. Division by the channel cross-section yields a linear flow speed of 10,000 cm h^{-1} = 2.8 cm s^{-1}. Thus, each fluid volume element spends about 1 s within the 3-cm long microchannel. During this time, volume elements at the co-flowing lines of the inner and outer phase are in contact, allowing the hemoglobin proteins to diffuse from the middle into the surrounding outer streams.

As a third step, to estimate the distance that a hemoglobin globule can diffuse normal to the flow direction within the fluid contact time of 1 s, we use the Einstein–Smoluchowski equation:

$$\sqrt{\langle x^2 \rangle} = \sqrt{2Dt} = \sqrt{2 \times 98 \mu\text{m}^2 \text{s}^{-1} \times 1 \text{s}} = \sqrt{196 \mu\text{m}^2} = 14 \mu\text{m}$$

This means the proteins penetrate into the outer streams (each of them occupying one-third of the 100-μm wide channel) to an extent of about 50% each. This is quite marked, and the laminar co-flowing streamlines will be notably smeared as the fluids flow further down the channel.

Further note: all the above is valid only in the (usual) case of no severe density mismatch of the fluids involved. If, by contrast, these fluids exhibit marked differences in their densities, then gravitational effects will play a marked role, leading to buoyant flow and mixing, where the denser fluids "crawls" underneath the less dense one in the channel. Yet again, there is a dimensionless number quantifying the interplay of buoyancy-driven flow v_{bf} and diffusion via the diffusion coefficient D, called the Rayleigh number Ra:

> **Rayleigh number Ra**
>
> **Interplay of buoyant flow (also known as free convection induced by density differences in the fluid system) and diffusion**
>
> $$Ra = \frac{\text{free convection}}{\text{diffusion}} = \frac{v_{bf} L_0}{D}$$
>
> v_{bf} Gravity-driven buoyant flow (m s^{-1})
> L_0 Characteristic channel size over which the interplay of convection and diffusion shall be appraised; in microfluidics, this is usually the channel diameter (m)
> D Diffusion coefficient (m^2 s^{-1})

> **Back of an envelope calculation**
>
> How long does it take for sugar ($D \approx 1000$ μm^2 s^{-1} in water at room temperature) to equally spread within an espresso mug (typ. height: 7 cm) without stirring, i.e., by diffusion only?
>
> Again, we use the Einstein–Smoluchowski equation, now rearranged for the time t:
>
> $$t = \frac{\langle x^2 \rangle}{2D} = \frac{(70{,}000\,\mu m)^2}{2 \times 1{,}000\,\mu m^2 s^{-1}} = 2{,}450{,}000\,s \approx 680\,h \approx 28\,d$$
>
> ...about one month!

We have seen from the examples above that diffusion is a relevant mechanism of fluid intermixing in microfluidics; it is, however, still a slow one. In many situations of microfluidic analytics, mixing shall be enhanced further, for example, if it is desired to probe the reaction rather than the diffusion kinetics of two components that are brought into touch in a microchannel. A key necessity for this endeavor is

FEM simulation of microfluidics
2. Diffusion in liquid flow (Part 1)

While fluid flow per se can be modeled by applying the Navier–Stokes equations, **mixing and reaction of solutes within and across fluid streams requires an additional PDE**, which is added to a simulation modeling of such fluid flow in microfluidic devices in the following fashon:

$$\delta_{ts}\frac{\partial c_i}{\partial t} = D_i \nabla^2 c_i + R_i(c_i, t)$$

δ_{ts}	**Time-scaling coefficient** (usually "1" for seconds, 1/60 for minutes)
c_i	**Concentration** of species i (mol m^{-3})
t	**Time** (s)
∇	**Nabla operator**, in a 3D Cartesian coordinate system $\nabla = \left(\frac{\partial}{\partial x}, \frac{\partial}{\partial y}, \frac{\partial}{\partial z}\right)$
D	**Diffusion coefficient** (m^2 s^{-1})
R	**Reaction rate** (mol m^{-3} s^{-1})

A typical example includes the mixing of a drug-loaded fluid stream being flow-focused into thin jet by a liquid stream that is a non-solvent for the drug in a microfluidic cross-junction. This leads to **diffusive mixing across the liquid–liquid interface**, a drop in solvent quality, and **nucleation of nanoparticles** of the respective drug.

FEM simulation of microfluidics
2. Diffusion in liquid flow (Part 2)

Post processing and visualization:

Simulation model combining PDEs for fluid flow (Navier–Stokes) and diffusion.

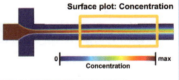
Surface plot: Concentration

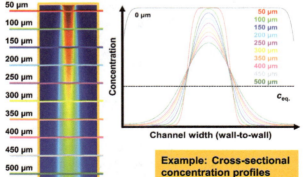

Example: Cross-sectional concentration profiles

Further reading:
S. Soh et al.: "Reaction-Diffusion Systems in Intracellular Molecular Transport and Control", *Angew. Chem. Int. Ed.* 2010, *49*, 4170. DOI: **10.1002/anie.200905513**.

to shorten diffusive path lengths as far as possible, such to reduce the times required to pass these distances. If that way of achieving mixing is not even enough, various further sophisticated forms of mixing-channel geometries have been engineered, some of which introduced at the end of Section 3.1.3. To predict complete mixing between two species in simple co-flow or complex microchannel geometries (cf. Section 2.1) as well as to appraise the diffusive path length that is required for complete mixing in the controlled environment of a microfluidic device, CFD simulations are again a valuable tool. For example, to model diffusion at the liquid boundary of two flowing streams, one can utilize a reaction-diffusion equation, as detailed in the two info boxes on FEM simulation of microfluidics 2.

> **Take-home message**
>
> **Diffusion is largely negligible on macroscopic scales, but it is marked on microscopic scales as typically used in microfluidics.**
>
> Table 1.1 in Section 1.1.2. compiles a list of viscosities of fluids typically used in microfluidics. With these, further estimates can be performed if the size of the diffusing molecules or particles of interest is known, which may be found in databank-literature or can be estimated / calculated by computer modeling.

1.4 Interfacial and capillary forces

1.4.1 Interfacial tension

To this end, we have established a set of concepts and equations to describe the motion of fluids based on directed advection and random diffusion. In both these realms of transport, we have chosen continuum approaches for the modeling, and it was necessary to consider the molecular or particulate composition of the fluids only to *illustratively* understand their motion. In all this preceding treatment, repeatedly, we have conducted balancing over selected volume- or area-elements in the fluid, but in all these instances, we have (silently) assumed that these are actually placed in the bulk of the fluid. As we will see now, the situation can be markedly different if we in fact consider those regions in space where the fluid phase ends, that is, if we consider the fluid **interface** to another medium.

We start by considering a liquid in a shallow vessel such as a Petri dish with an air atmosphere above. If we carefully place a light object like a thin needle on the liquid surface, we observe that it swims on top of it, as sketched in Figure 1.17. This is not due to buoyancy, but instead, due to **surface tension**. The molecules in the liquid exert forces to one another in the form of non-covalent interactions such as van-der-Waals forces, dipole–dipole interactions, or hydrogen bonds (the latter are of prime relevance in water). For each molecule in the bulk of the liquid, these forces are isotropically balanced in a time average, because each molecule is isotropically surrounded by neighbors, as shown in Figure 1.18. Molecules on the surface, by contrast, are surrounded only by those underneath or aside of them, whereas above of them are molecules of another kind (for example, those of a gas atmosphere such as air); as a result, these surficial molecules are subject to an *anisotropic* environment of forces, as also sketched in Figure 1.18. If a surface molecule is lifted up, the non-isotropically balanced bonds to its alike neighbors below are stretched, which gives rise to a restoring force acting to pull the molecule back to its equilibrium position. Vice versa, if a lightweight object like a needle is placed on the liquid surface, molecules on the surface will be pushed down, and the non-covalent bonds to their neighbors below will be compressed; yet again, this effect gives rise to a restoring force pushing the molecules back to their equilibrium positions, and as long as the gravitational force of the needle does not outweigh this back-pushing force, the needle can swim on the liquid surface. We may therefore imagine the liquid surface to be under tension like a membrane – referred to as surface tension.

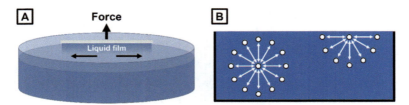

Figure 1.17: Consequences and origin of surface tension of a liquid in an open vessel. (A) Film formation between the surface of a liquid and a thin long object (e.g., a needle) being pulled out by a force *F* normal to the liquid surface. The surface tension gives rise to emergence of a two-sided liquid film pinned to the needle that is pulled out of the liquid into the direction of *F*. (B) Schematic of mutual forces acting between a liquid molecule and its neighbors in the bulk and at a liquid surface. In the bulk, forces act on a molecule isotropically such that the net force is zero (left). At the surface, by contrast, the anisotropic environment of forces gives rise to a nonzero net force counteracting displacement of the surface molecules in the normal direction (right).

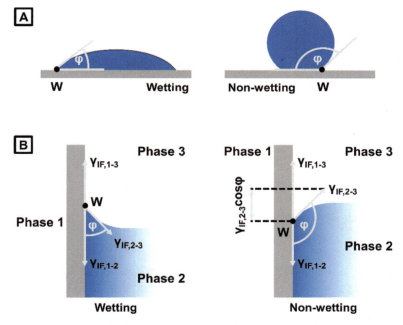

Figure 1.18: Wetting (left) and non-wetting (right) of a liquid on (A) a horizontal solid surface and (B) at a vertical wall, such as that of a capillary. At the line of contact of the solid material (denoted 1), the liquid (denoted 2), and the gas (most typically air) (denoted 3), commonly named wetting line and therefore labeled with W in the sketches, three interfacial tensions are operative: $\gamma_{IF,1-2}$, $\gamma_{IF,1-3}$, and $\gamma_{IF,2-3}$, all acting tangentially to the respective interfaces. At equilibrium, the sum of the resulting forces is zero, resulting in a contact angle. This angle is small for a wetting liquid (left-side sketches), whereas it is large for a non-wetting liquid (right-side sketches).

If we carefully pull the needle away from the liquid surface in Figure 1.17, a thin film of the liquid will be pinned to it and stretched out of the liquid. The force that is necessary to achieve that is proportional to the extension of the liquid surface that comes along with this stretching, plus the trivial increment $m\mathbf{g}$ needed to lift the needle itself:

$$F = 2\gamma_{IF} l \mathbf{e}_z + m\mathbf{g} \tag{1.59}$$

with $\mathbf{e}_z$ the unit vector in the direction of the stretching (here: z), which is the one normal to the (flat) liquid surface, and γ_{IF} the interfacial tension of the liquid to the adjacent medium under consideration. In the present scenario, we consider an interface of a liquid to a gas (here: air), such that the interfacial tension is referred to by its special-case name surface tension. The factor 2 in eq. 1.59 emerges because the liquid film that is stretched in Figure 1.17 has two surfaces, one on its front and one on its backside.

> **Note**
>
> It is unfortunate that the letter γ is used for both describing the interfacial tension and the extent of shear strain in the context of fluid physics, which is not our invention. In the context of this textbook, shear itself is not of prime relevance, but only its time-derivative, which appears in the fundamental Newton law:
>
> $$\sigma = \eta \frac{\partial \gamma}{\partial t} = \eta \dot{\gamma}$$
>
> Hence, whenever $\partial \gamma / \partial t$ is used in this textbook, it refers to the shear rate, whereas the symbol γ_{IF} refers to interfacial tension.

In the view of eq. 1.59, the interfacial (or surface) tension quantifies the *force* necessary to extend the liquid interface (or surface) by a unit increment. In this view, the physical dimension of this quantity is N m^{-1}. In an alternative view, we may consider the *work* required to extend the surface of a medium, A, by a unit amount ∂A, which again is just nothing else than γ_{IF}:

$$\partial w = \gamma_{IF} \partial A \qquad (1.60)$$

Related to that latter viewpoint, in chemical thermodynamics, the surface or interfacial tension is commonly introduced as the partial derivative of the Gibb's free energy by area; that is, it quantifies the change of the liquid's free *energy* if the surface changes:

$$dG = \gamma_{IF} dA \qquad (1.61)$$

In the view of eqs. 1.60 and 1.61, the interfacial (or surface) tension has an energy-related meaning, which reflects the macroscopic effect of the microscopic intermolecular interaction energies; in this view, the alternative suitable physical dimension of this quantity is J m^{-2}, which is equal to N m^{-1}. Equation 1.61 shows that the interfacial tension, γ_{IF}, must be positive, because otherwise, the free energy of any multiphase system could decrease by increase of the interfacial area, which would mean that stable interfaces could not exist at all, but by contrast, different phases would tend to intermix as much as possible such to maximize their mutual interfacial contact. Our knowledge on multiphase systems tells us that this is not the case.

Based on this premise, eq. 1.61 then in turn shows that a system can minimize its Gibb's free energy by minimizing its surface; this is the reason why liquid droplets or gas bubbles are commonly spherical, unless they are subject to further forces such as flow fields in microfluidic channels or gravity. Another very important type of such further forces are those that arise from interactions with yet further adjacent materials, as discussed in the following section. Before considering this aspect in further detail, we re-emphasize what we have just stated: a liquid droplet or gas bubble suspended in an external immiscible medium can minimize its free energy by minimizing its interface to the external medium. As quantified by eq. 1.61, a decrease of the droplet or bubble radius, R, comes along with a gain in free energy of:

$$\Delta G = \gamma_{IF} \Delta A = \gamma_{IF} 4\pi (R_{before}^2 - R_{after}^2) = \gamma_{IF} 4\pi \left[R^2 - (R - \Delta R)^2 \right] \quad (1.62)$$

This process, in turn, leads to an increase of the inner pressure of the droplet or bubble, Δp, requiring work of amount $w = \Delta p \Delta V = \Delta p\, 4\pi R^2 \Delta R$ to be performed. Balancing ΔG and w yields:

$$\gamma_{IF} 4\pi \left[R^2 - (R - \Delta R)^2 \right] = \Delta p\, 4 p R^2 \Delta R \quad (1.63)$$

Rearranging this expression for Δp and neglecting terms with ΔR^2 yields the **Laplace pressure**:

$$\Delta p = \frac{2\gamma_{IF}}{R^2} \quad (1.64)$$

The smaller the droplet or bubble radius, R, the larger is this excess pressure. This circumstance means that if droplets or bubbles are in contact with one another, large ones can reduce their excess pressure by growing on cost of small ones. As a result, whenever droplets or bubbles of a dispersed component in a liquid–liquid emulsion or a gas–liquid foam have a finite size dispersity along with a finite solubility of the dispersed component in the continuous phase, larger droplets or bubbles of it will incorporate material from the smaller ones by diffusion through the continuous phase. Thereby they grow on cost of the smaller ones, which will eventually fully break the emulsion or the foam. This process is named **Ostwald ripening** [57]. To prevent that from happening, a substance with absolute no solubility in the continuous phase must be added to the dispersed phase such to contribute an additional osmotic pressure π_{Osm} that counteracts the Ostwald ripening.

1.4.2 Cohesive versus adhesive forces: Wetting and capillarity

Figure 1.17 illustrates the forces between a molecule and its alike neighbors in a liquid medium. These forces are named **cohesive forces**. In addition to these cohesive forces, a molecule of the liquid can also undergo interactions with other substances in the system, for example, with the walls of the container or with the walls of a microchannel in the case of a microfluidic experiment. These forces are named **adhesive forces**. If the adhesive forces outweigh the cohesive forces, which is the case, for example, for water in a glass microcapillary, the liquid is termed to be **wetting**. In this case, a droplet of the liquid tends to spread on a surface of the other material, as shown in Figure 1.18A (top schematic). Alternatively, if the liquid finds itself in a tiny capillary of the wettable material, the surface of the liquid is bend in a concave fashion, as also shown in Figure 1.18A (lower schematic). The **contact angle**, φ, between the wall and the liquid surface is a measure of the relative strength of the cohesive and adhesive forces. For a liquid that wets a surface, we find $\varphi < 90°$. By contrast, if the cohesive forces in the liquid outweigh its adhesive forces with another material, which is the case, for example, for mercury and glass, the liquid does not wet the surface. In this scenario, a droplet of the liquid tends to roll-up in an almost perfectly spherical shape such to minimize interfacial contact with the other medium, as shown in Figure 1.18B (top schematic). Alternatively, if that kind of liquid finds itself in a tiny capillary of the non-wettable material, the surface of the fluid is bend in a convex fashion; the contact angle is then $\varphi > 90°$.

To fully quantify the preceding view, we must consider three interfacial tensions and their effects in the experimental situations sketched in Figure 1.18, which are those between the solid capillary material (denoted 1), the liquid (denoted 2), and the gas (most typically air) on top (denoted 3). Between these three phases, there are three interfacial tensions, $\gamma_{IF,1-2}$, $\gamma_{IF,1-3}$, and $\gamma_{IF,2-3}$, acting tangentially to the respective interfaces. If we consider a line element, dl, which is oriented perpendicular to the paper plane in the region of contact of the three phases (spot W in Figure 1.14, commonly named *wetting line*), we can appraise there to be a force ($\gamma_{IF,1-2} - \gamma_{IF,1-3}$) dl parallel to the solid surface and a force $\gamma_{IF,2-3}\, dl$ parallel to the liquid surface, which projects a component $\gamma_{IF,2-3} \cos\varphi\, dl$ to the solid-surface direction. At equilibrium, the sum of these forces is zero:

$$\gamma_{IF,1-2}dl - \gamma_{IF,1-3}dl + \gamma_{IF,2-3}\cos\varphi\, dl = 0 \Leftrightarrow \frac{\gamma_{IF,1-3} - \gamma_{IF,1-2}}{\gamma_{IF,2-3}} = \cos\varphi \qquad (1.65)$$

The latter eq. 1.65 is termed Young's equation [58] it further substantiates our above understanding: if the solid–liquid interfacial tension, $\gamma_{IF,1-2}$, is smaller than the solid–gas interfacial tension, $\gamma_{IF,1-3}$, it is favorable to replace solid–gas contact by solid–liquid contact, which is achieved if the liquid in the capillary exhibits up-climbing at the walls such to obtain a concave surface; in this case, $\cos\varphi > 0$ according to eq. 1.65, such that $\varphi < 90°$. A prime example is water on glass, with a contact angle of almost 0°. Vice versa, if the solid–liquid interfacial tension, $\gamma_{IF,1-2}$, is larger than the solid–gas interfacial tension, $\gamma_{IF,1-3}$, it is favorable to replace solid–liquid contact by solid–gas contact, which is achieved if the liquid in the capillary climbs down the walls such to obtain a convex surface; in this case, $\cos \varphi < 0$ according to eq. 1.65, such that $\varphi > 90$. A prime example is mercury on glass, with a contact angle of about 140°.

We may further follow the above line of thought to understand yet another effect observed on liquids in capillaries: in the case of a wetting liquid, which is the situation sketched in Figure 1.18A, the surface tension at the wall has an up-pointing component, as we have just quantified above. As a result, the liquid will climb up the capillary until the surface-tension-based uplifting force is balanced by the downward gravitational force of the liquid volume element. This effect is named **capillary effect**. We can quantify its extent by considering a capillary of radius r and a liquid element to climb up a height h, as sketched in Figure 1.19. The force that causes the climbing is the vertical component of the surface tension, which is $\gamma_{IF} \, l \cos \varphi$ according to eq. 1.59. Note that the factor 2 of eq. 1.59 does no longer appear here, as we do not consider a liquid film with two sides. The length l equals the circular capillary circumference $2\pi r$, such that we get $2\gamma_{IF}\pi r \cos \varphi$. The counteracting gravitational force is $m\mathbf{g} = r(\rho r^2 h)\mathbf{g}$. Balancing these two forces yields an expression for the climbing height, $h = 2\gamma_{IF} \cos \varphi \, / \, \rho r \mathbf{g}$, which provides an easy way to experimentally measure the surface tension. In the opposite case of a nonwetting liquid, which is the situation sketched in Figure 1.18B, the liquid surface has a convex curvature, causing a *downward* force that drives the liquid element in the capillary to climb down, which is an effect named *capillary depression*.

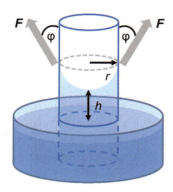

Figure 1.19: Climb-up of a liquid in a capillary made of a well-wetting material.

In microfluidic channels that shall serve to flow multiple streams of fluids with different polarities, in many experimental situations, it is crucial to ensure good wettability of certain of these fluids, whereas others shall have no good wettability in the given situation [59], as will be discussed in greater detail in Section 3.4.1. To realize that, the microchannels must be treated with appropriate surface modifiers, as also discussed in greater detail in Section 3.4.1. The contact angle φ, as introduced in this section, is a useful parameter to quantify the performance of such a surface modification procedure prior to conducting a microfluidic experiment. For this purpose, the same surface modification process is applied to a flat surface of the microchannel material (often glass or silicone), and then a drop of the fluid of interest is placed onto it. If it spreads on the surface and exhibits a small contact angle, as shown in the left sketch of Figure 1.18A, there is good wettability, whereas if the droplet furls to adopt a spherical shape with a large contact angle to the flat surface, there is poor wettability.

1.4.3 Interfacially active substances: Surfactants

Up to this point, we have treated interfacial effects of pure liquids that consist of one component only. We now drop this constraint and consider multicomponent systems such as water-in-oil or oil-in-water emulsions. As a starting point, we regard the fundamental thermodynamic Gibbs–Duhem equation in a system with i components at isobaric and isothermal conditions:

$$dG = A d\gamma_{IF} + \sum n_i d\mu_i \qquad (1.66)$$

with n_i the molar amount and μ_i the chemical potential of the i-th component in the system, whereas A denotes the area of the dividing interface. At equilibrium, we have $dG = 0$; rearranging eq. 1.66 then yields:

$$d\gamma_{IF} = -\sum \frac{n_i}{A} d\mu_i = -\sum \Gamma_i d\mu_i \qquad (1.67)$$

Here, the ratio of n_i / A is named **surface excess**, Γ_i; it quantifies the excess number of molecules of component i that sit at the interface of a multicomponent phase. If a component i consists of molecules that favorably localize in that interface, it is named interfacially active or surface active agent, in short **surfactant**. Such molecules are amphiphilic; they contain both a hydrophilic part, often named head group, and a hydrophobic part, often named tail. We can generally distinguish between four classes of surfactants, as illustrated in Figure 1.20. Anionic surfactants contain a hydrophilic, negatively charged head group (e.g., COO^-, SO_3^{2-}, SO_4^{2-}), cationic surfactants mostly contain a quaternary ammonium group, also named "quat," amphoteric surfactants comprise a both negatively and positively charged head group, and nonionic surfactants do not contain any charge in their head group, but instead, hydroxyl, ether, or amide groups. In typical examples, the tail of surfactants is an either straight or branched hydrocarbon or fluorocarbon chain of commonly 8 to 18 carbon atoms.

Whereas the head of surfactants favorably interacts with polar media, the tail favorably does so with unpolar media. This amphiphilic constitution drives surfactant

Figure 1.20: Four main classes of surfactants, distinguished from one another by their type of head group.

molecules to the interphase of a polar and an unpolar phase, where they can arrange themselves such that their polar heads stick into the polar phase, whereas their unpolar tails point into the unpolar phase, as sketched in Figure 1.21. In this figure, we consider a particularly relevant kind of two-phase systems for microfluidics: **emulsions**. In an emulsion, one fluid is dispersed within another fluid of opposing polarity in the form of droplets. This process generates a thermodynamically out-of-equilibrium system, as the formation of additional interfacial area is costly in terms of energy. This implies that the minimum energy of a two-liquid system corresponds to a fully separated two-phase system. Therefore, if no surfactant is present, droplets naturally minimize their interfacial areas, and with that, minimize the system's free energy according to eq. 1.61, by undergoing **coalescence**. To prevent coalescence, surfactants can take action: their specific arrangement at the interface leads to stabilization of the droplets of the dispersed liquid in the continuous liquid. For example, if oil is the inner and water is the outer fluid, which is the situation sketched in Figure 1.21, the polar head groups of a surfactant that stick out of the oil droplets will electrostatically repel each other in the continuous water phase, thereby preventing the oil droplets to pervade into one another and coalesce. In the other case of water-in-oil droplets, the hydrophobic tails of the surfactant will stick out at the droplet interface and repel one another for steric reasons, again preventing the droplets to pervade into one another and coalesce.

As a result of their enrichment at the interface, surfactants have large positive surface excess, Γ_S. By contrast, common liquid phases in microfluidics such as oil or water have no tendency for enrichment at (or depletion from) their mutual interface, such that their surface excess is usually zero. With that, eq. 1.67 adopts a simplified form:

$$d\gamma_{IF} = -\Gamma_S d\mu_S \qquad (1.67a)$$

The concentration dependence of the chemical potential is $d\mu_S = RT\, d(\ln c_S)$ (cf. eq. 1.36), with the molar concentration c_S of the surfactant in the bulk phase. This yields $d\gamma_{IF} = -RT\, \Gamma_S\, d(\ln c_S)$. Applying the identity $d(\ln c_S) = (1/c_S)\, dc_S$, we get the Gibbs adsorption isotherm for dilute solutions [60]:

$$\left(\frac{d\gamma_{IF}}{dc_S}\right) = -\frac{RT\Gamma_S}{c_S} \qquad (1.68)$$

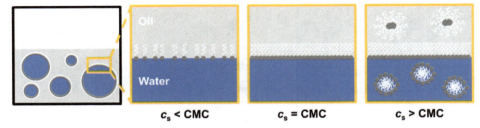

Figure 1.21: Schematic of a water-(blue color)-in-oil (gray color) emulsion. Droplet coalescence is prevented by a surfactant that sits at the oil–water interphase at an arrangement where its unpolar tail sticks into the oil phase, whereas its polar head group stick into the water phase, as sketched in the three zoom-in schematics. This arrangement of the surfactant at the interface lowers the interfacial tension and stabilizes the emulsion from breaking by spontaneous droplet coalescence. In the first zoom-in schematic, the surfactant overall concentration, c_S, is below the critical micelle concentration, CMC, such that the surfactant can only partly populate the interface. In the middle zoom-in schematic, the surfactant concentration exactly matches the CMC, such that the surfactant does fully cover the interface. In the right zoom-in schematic, the surfactant is present at a concentration larger than the CMC, so that in addition to fully covering the interface, excess surfactant will arrange itself in the form of micelles in the bulk of either or (as shown here) both phases (with a partition that depends on its bulk solubility in the two phases). At surfactant concentrations not too far above the CMC, these micelles will be spherical, as sketched in this figure, whereas at yet higher surfactant concentration, various other forms of micelles can form.

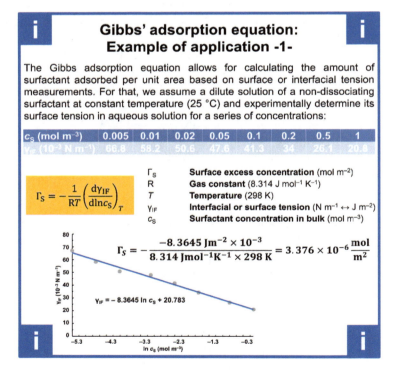

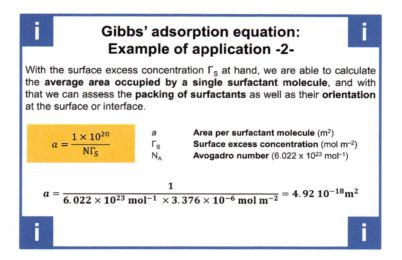

The latter equation gives a quantitative view on the action of surfactants: as they favorably localize at the phase interface, they contribute a large positive Γ_S, such that according to eq. 1.68, $(d\gamma_{IF}/dc_S)$ will be negative. As a result, adding surfactant will *reduce the interfacial tension* of liquid two-phase systems such as emulsions. This holds up to a point where the concentration of a surfactant gets larger than a critical value corresponding to such an amount that all interfacial area is just fully covered by surfactant molecules. From this point on, addition of further surfactant will cause excess surfactant to be present in the bulk of either one or both of the liquid phases. These excess molecules of surfactant will arrange themselves in the form of **micelles**, as sketched in Figure 1.21 (right window). The critical concentration whereupon this happens is named **critical micelle concentration**, CMC. From the CMC on, further addition of surfactant will no longer reduce the interfacial tension, as visualized in Figure 1.22. The slope of the linear part of the curve in Figure 1.22 right below the CMC allows the surface excess, Γ_S, to be determined by eq. 1.68. The simplest form of micelles, which is present at concentrations just above the CMC, is spherical micelles, as illustrated in Figure 1.21. Further increase of the surfactant concentration, in further interplay with temperature, can lead to various other forms of micelles, as discussed in further detail elsewhere [61].

The physical principles and equations discussed in this section inherently contain means to *quantify* interfacial tensions, which is necessary if microfluidic flows shall be quantified and modeled on basis of the principles detailed in Section 1.1. The following info boxes detail somewhat further on such experimental means to quantify interface and surface tensions.

Measuring interfacial tension (Part 1)

A theoretically simple but experimentally puzzling means of quantifying the surface tension of a liquid is based on equation 1.59:

$$F = 2\gamma_i l e_z + m g$$

If an object such as a needle or, as a common variant, a ring of wire is lifted and detached from the surface of the liquid of interest, a pending liquid film gives rise to an interfacial-tension-based force that adds up to the inherent gravitational force of the object:

Linear needle with length l
$$|F| = 2\gamma_{IF} l_z + m|g|$$

Ring-shaped wire with radius r
$$|F| = 2 \times 2\pi r \gamma_{IF} + m|g|$$

(Again, note that there is a factor 2 in both these equations accounting for the two surfaces of the pending liquid film, which has two sides.)

Measuring interfacial tension (Part 2)

Whereas the preceding method is most suitable to measure *surface* tensions, quantification of *interfacial* tensions in a liquid–liquid two-phase system is even more relevant in the area of microfluidics. A suitable approach is based on pending a droplet of the one fluid from a capillary immersed within a continuous phase of the other liquid.

At equilibrium, the drop exhibits a shape profile that is a function of the capillary tube radius, the interfacial tension γ_{IF}, and the density difference of the two liquids:

$$\gamma_{IF} = \frac{\Delta\rho |g| d_{eq}^2}{H}$$

d_{eq}: droplet equatorial diameter
d_{neck}: droplet diameter at its neck
H: function of d_{eq}/d_{neck}

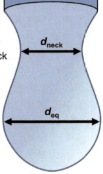

Further reading:

J. M. Andreas et al.: "Boundary Tension by Pendant Drops", *J. Phys. Chem.* 1938, *42*, 1001. **DOI: 10.1021/j100903a002.**

C. E. Stauffer, "The Measurement of Surface Tension by the Pendant Drop Technique", *J. Phys. Chem.* 1965, *69*, 1933. **DOI: 10.1021/j100890a024.**

A. Nakajima et al.: "A Pendant-Drop-Type Tensiometer for Surface Tension Measurements of Polymer Systems", *Polymer J.* 1972, *3*, 640. **DOI: 10.1295/polymj.3.640.**

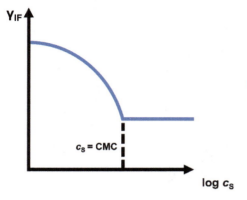

Figure 1.22: Interfacial tension γ_{IF} depending on the logarithmic surfactant concentration, log c_S. In the linear domain right below the CMC, equation 1.68 holds and can be applied to estimate the amount of excess surfactant at the interface, Γ_S. Adapted from T. F. Tadros, "An Introduction to Surfactants", De Gruyter, Berlin (2014).

1.5 Microfluidic flow patterns

1.5.1 Co-flow and segmented flow

In Section 1.1.3., we have introduced the striking inertial irrelevance in microfluidics, which manifests itself in the form of inherently small Reynolds numbers. This causes fluids to idyllically flow alongside one another, as sketched in Figure 1.23A; it is, however, only valid if the fluids are *miscible*, that is, if there is no interfacial tension between them. For instance, we can inject ethanol into a flow-focusing junction and form a continuous jet of it by co-injection of water (cf. schematic in Figure 1.23A and Section 3.3, for example, Figure 3.27 in there). This situation is named **co-flow**. If, by contrast, there is a distinct interfacial tension present between two fluids, microfluidics leads to **segmented flow**, creating droplet emulsions, as sketched in Figure 1.23B. This is why segmented flow in microfluidics is often called droplet-based or droplet microfluidics. These are **two fundamentally different modes of operation of microfluidics devices**. By suitable combination of them, a rich variety of flow patterns can be realized. For example, combining a co-flow and a segmented-flow element may serve to form droplets with a controlled anisotropic composition such as the famous variant of "Janus droplets" shown in Figure 1.23C, whereas multiple (often: double) sequential segmented flow allows complex emulsions such as double emulsions – drops that carry smaller drops inside – to be formed, as illustrated in Figure 1.22D. All this is detailed further in later sections of this book.

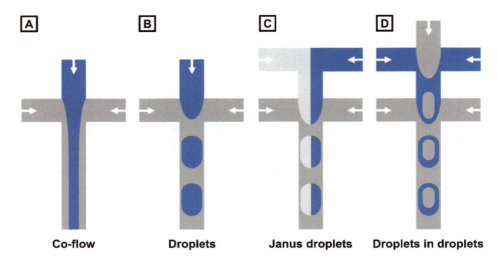

Figure 1.23: Variants of co-flow and segmented-flow microfluidics, schematized by top-view sketches of microfluidic channels with one or two rectangular junctions wherein different fluids are mixed to flow downstream with one another. (A) Laminar co-flow of two miscible fluids. (B) Droplet formation by flow-focusing of an inner fluid exerted by an outer immiscible fluid. (C) Combination of (A) and (B): In the upper part of the microflow cell, two miscible fluids co-flow in a laminar stream, which is then periodically broken to form Janus-shaped droplets by flow-focusing with a third, immiscible fluid. (D) Extension of method (B): double-sequential flow-focusing with three mutually immiscible fluids leads to formation of droplets inside droplets, referred to as double emulsions.

1.5.2 Segmented flow: Viscous versus interfacial drag

In a typical segmented-flow experiment, immiscible fluids such as water and oil are flown to meet at a microchannel junction. In the first early reports on this matter [62], the junction geometry was a T-shaped channel patterned into a suitable material such as glass or silicone, as shown in Figure 1.24A.

In such a junction, competing stresses drive the interface: interfacial tension acts to reduce the interfacial area, whereas viscous stresses act to extend and drag the interface downstream. These competing stresses destabilize the interface and cause **droplets** with radius R to form [63]. The interplay of the viscous and the capillary-interfacial forces is captured by the dimensionless **capillary number**.

The principle of the latter type of drop formation can be illustratively explained using a water faucet as an example. If we turn on a faucet at a low flow rate, water drips out one drop at a time. The drop size is a result of a balance between the surface forces of the hanging drop and gravity, as illustrated in Figure 1.25, and therefore depends on the surface tension of the fluid and the size of the faucet. Since both the surface tension and the faucet size are constant, all drops dripping from a faucet exhibit a narrow size distribution. In a microfluidic device, a similar

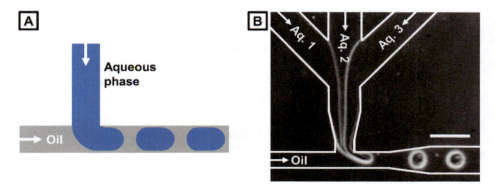

Figure 1.24: T-shaped microchannel geometries for forming emulsion droplets. (A) Schematic of formation of a water-in-oil (W/O) emulsion in a T-junction. Interface formation between the two immiscible fluids is driven by competing stresses: viscous shear stresses tend to extend and drag the interface, whereas surface tension tends to reduce the interfacial area. This competition leads to a droplet size of order $R \sim Ca^{-1}$. (B) Microdroplets formed from three different aqueous phases in a T-junction. The scale bar denotes 100 µm.

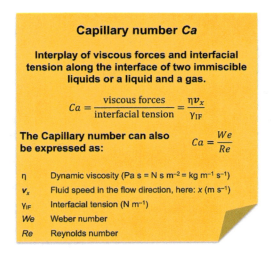

Capillary number Ca

Interplay of viscous forces and interfacial tension along the interface of two immiscible liquids or a liquid and a gas.

$$Ca = \frac{\text{viscous forces}}{\text{interfacial tension}} = \frac{\eta v_x}{\gamma_{IF}}$$

The Capillary number can also be expressed as:

$$Ca = \frac{We}{Re}$$

η	Dynamic viscosity (Pa s = N s m^{-2} = kg m^{-1} s^{-1})
v_x	Fluid speed in the flow direction, here: x (m s^{-1})
γ_{IF}	Interfacial tension (N m^{-1})
We	Weber number
Re	Reynolds number

interplay is operative. Just like a faucet, when operated at a low flow rate, a microfluidic device can produce fluid droplets one at a time. There are largely three differences to the faucet case. First, in a microfluidic system, the external phase is not air like in our faucet example, but a flowing liquid in a microchannel; this fluid necessarily contains a small portion (typically approximately 1%, w/w) of a surfactant to stabilize the droplets that form from undergoing coalescence. Second, whereas a faucet operates at droplet production rates up to just a few Hertz, a microfluidic dropmaker can produce droplets at rates much higher, up to kilohertz. Third, whereas a faucet produces drops with sizes in the millimeter range, microfluidic devices do so in the range below, typically with tens to hundreds of micrometers in diameter. Just like in the faucet analogy, the droplet size in a microfluidic system is a result of a balance between the interfacial forces of the pending drop and the shear exerted by the continuous fluid phase (Figure 1.25). Therefore, the size of the droplets depends on the interfacial tension of the fluid, the flow rates in the channel, and the size of the microchannel constriction in the region of drop formation. In a quantitative assessment, capillary stresses of magnitude $\gamma_{IF}/R_{Droplet}$ are balanced by viscous stresses of magnitude $\eta v_x/h$, where h is the height of the microchannel, thereby giving a nearly monodisperse droplet size.

$$R_{Droplet} = \frac{\gamma_{IF} h}{\eta |v_x|} = \frac{h}{Ca} \qquad (1.69)$$

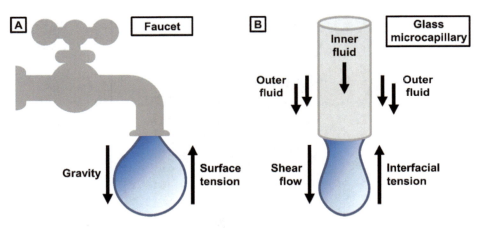

Figure 1.25: Analogy of droplet formation in (A) a dripping faucet and (B) a heterophase microfluidic system in the dripping regime, both operated at low flow rates. (A) The faucet produces one droplet at a time based on a force balance between surface tension acting to keep the drop pinned to the faucet and gravity acting to pull the drop down; once the latter force prevails, the droplet pinches off. (B) A similar interplay of forces is operative in a microfluidic system, where interfacial tension is overcome by external shear flow in the moment of droplet pinch-off.

The schematic in Figure 1.25B illustrates a simple but powerful alternative to the T-junction channel geometry in Figure 1.24, which is that of co-axially aligned glass microcapillaries. In this variant of a microfluidic device, a microcapillary with a square cross-section co-axially nests another microcapillary with a round cross-section and a tapered end, as sketched in more detail in Figure 1.26A. Two immiscible fluids are flown through these in the same direction, an inner fluid flowing through the inner round-section capillary and an outer fluid flowing through the interstitial volumes between the inner round capillary and the outer square capillary. When both fluids flow at low rates, individual monodisperse drops are formed periodically at the tip of the capillary orifice in a process termed dripping, as shown in Figure 1.26B [64, 65]. In this dripping regime, just as in the related one shown in the T-junction device in Figure 1.24A, the drop at the end of the capillary tube experiences two competing forces: viscous drag pulling it downstream and forces due to interfacial tension holding it to the capillary. Initially, interfacial tension dominates, but as the attached droplet grows, drag forces eventually become comparable and then prevail; this is when the droplet breaks off and is carried away by the flow of the continuous phase. If one of the two fluids flows at a much greater flow rate than the other, by contrast, the droplet pinch-off occurs further downstream by one of two sorts of jetting mechanisms, as shown in Figure 1.26C and 1.26D and detailed further below.

An alternative geometry for drop formation in capillary devices is the flow-focusing geometry [66], as shown in Figure 1.26E. In contrast to co-flow capillary devices, the two fluids are introduced from the two ends of the same square capillary, from opposite directions. Upon meeting at the tip of the tapered round capillary, the inner fluid is hydrodynamically focused by the outer fluid through the narrow orifice. At dripping conditions, drop formation occurs as soon as the inner fluid enters the circular orifice. An advantage of this method is that it allows us to make monodisperse drops with sizes smaller than that of the orifice; this feature is useful for making small droplets (typically approximately 10–50 μm in diameter). Flow-focusing drop formation can also be realized in microfluidic devices not made by assembling glass microcapillaries but by patterning channel structures into a suitable material such as glass or silicone elastomers [59, 67]. A common flow-focusing channel geometry in these devices is that of a four-way cross-junction, as shown in Figure 1.23B and 1.26F. In this geometry, the dispersed phase is injected into the central inlet and the continuous phase into the two side inlets. These fluids meet in the nozzle where, again, drops are formed.

1.5.3 Dripping and jetting

In all the microfluidic-device variants shown in Figure 1.23 to 1.26, if we increase the flow rate of either fluid beyond a certain critical limit, the result is a **jet**, which is a long stream of the inner fluid with drops forming downstream, as shown in Figure 1.26C and 1.26D. These drops have a broader size distribution than those formed from dripping, because the point at which the drops separate from the jet changes with each drop. At a deeper view, we can distinguish two **dripping-to-jetting transitions** [68, 69]. The first transition is driven by the flow rate of the outer fluid: as it is increased, drops formed at the tip decrease in size until the emerging fluid is finally

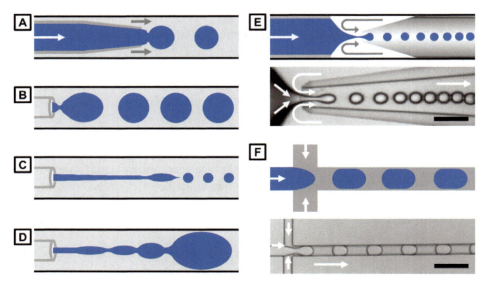

Figure 1.26: Formation of single emulsions in glass-capillary microfluidic devices with co-flow geometry as well as in glass-capillary and stamped microfluidic devices with flow-focusing geometry. (A) Schematic of the co-flow type of operation such to make droplets. Arrows indicate the flow direction of fluids and drops. (B) Schematic of drop formation at optimized flow rates of the disperse and the continuous phase, promoting formation of uniform droplets at the tip of the inner microcapillary (dripping regime). In contrast to that type of controlled droplet formation, jetting regimes lead to nonuniform droplet ensembles. There's two of such regimes. In one case, (C) a narrow jet is generated by increase of the flow rate of the continuous fluid above a threshold value. In another case, (D) a wide jet is generated by increase of the flow rate of the dispersed fluid above a threshold value. (E) Schematic and actual micrograph of a flow-focusing glass microcapillary device. (F) Schematic and actual micrograph of a flow-focusing microfluidic device with a four-way cross-junction channel geometry, prepared by combined photo- and soft lithography such to pattern these channels into a suitable elastomer material, most typically poly(dimethylsiloxane) (PDMS).

stretched into a jet. At this point, droplet breakup occurs downstream at the end of the thin jet, as shown in Figure 1.26C. The second transition is driven by the flow rate of the inner fluid: as it is increased, the dripping event is pushed downstream, and droplets ultimately pinch off from the end of the resultant jet. A similar effect to the latter can be observed in our example with a water faucet: if we gradually increase the flow rate through the faucet, a thin water stream or jet is formed. This jet eventually also breaks up into drops, but these have a larger size range and broader size distribution than those formed at low flow rates, where one drop at a time pinches off the faucet right at its tip. In both the faucet analogy and in the microfluidic situation shown in Figure 1.26D, the physics behind the breaking of the jet is the **Rayleigh–Plateau instability**, based on original observations made by Joseph Plateau [70] and Lord Rayleigh [71]. The driving force of this instability is that liquids, by virtue of their surface tensions, tend to minimize their surface area, and that there are tiny perturbations in a jet of liquid. If these perturbations are resolved into sinusoidal components, it is found that some grow with time while others decay. Among those that grow, some grow at faster rates than others, as determined by the components' wave numbers (number of peaks and troughs per centimeter) and the radius of the original cylindrical stream. As time proceeds, it is the component whose growth rate is maximum that will come to dominate and will eventually be the one that pinches the stream into drops.

Based on these observations, drop formation in microfluidics can be classified as being shear dominated or pressure dominated. Shear-dominated droplet formation tends to occur in unconfined geometries, in which the capillary number is near unity, $Ca \approx 1$, whereas pressure-dominated droplet formation occurs in confined geometries, in which $Ca < 0.01$. Yet, in a typical microfluidic experiment in the confinement of microchannels, as shown in Figure 1.18, is unity, $Ca > 0.01$. Droplet formation is thus likely a combination of shear and pressure effects. To further elucidate droplet formation, we consider the plot of a single-phase diagram based on the relevant balance of forces that induce the dripping-to-jetting transitions, as shown in Figure 1.27.

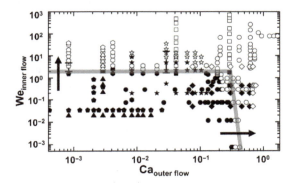

Figure 1.27: Dependence of the transition between dripping and jetting on the capillary number of the outer flow and the Weber number of the inner flow of a co-flow microcapillary device. Squares and diamonds: $\eta_{in}/\eta_{out} = 0.01$, with slightly different device geometries; hexagons and circles: $\eta_{in}/\eta_{out} = 0.1$, with slightly different device geometries; pentagons: $\eta_{in}/\eta_{out} = 1$; triangles, $\eta_{in}/\eta_{out} = 10$. Reprinted from [68]. Copyright 2007 American Physical Society.

We find that the behavior of droplet formation is determined by two dimensionless quantities. The **capillary number** relates the shear force on a liquid jet exerted by an outer continuous phase to its surface tension, whereas the **Weber number** reflects the balance between inertial forces of the inner liquid pushing the drop downstream and again the surface tension forces resisting the flow.

We also find, that for $\{Ca, We\} > 1$, the dispersed phase does not break into single droplets, whereas for $\{Ca, We\} < 1$, the liquid stream breaks into droplets due to a dripping instability. As a result, the boundary between dripping and jetting, as

> **Weber number *We***
>
> **Interplay of a fluid's inertial forces compared to its surface tension.**
>
> $$We = \frac{\text{interfacial forces}}{\text{surface tension}} = \frac{\rho v_x^2 L_0}{\gamma_{IF}}$$
>
> The Weber number can also be expressed as: $We = Ca \times Re$
>
> | | |
> |---|---|
> | ρ | Fluid density (kg m^{-3}) |
> | v_x | Fluid speed in the flow direction, here: x (m s^{-1}) |
> | L_0 | Characteristic length (e.g. droplet diameter) (m) |
> | γ_{IF} | Interfacial / Surface tension (N m^{-1}) |
> | Ca | Capillary number |
> | Re | Reynolds number |

shown in Figure 1.27, occurs when either number, or their sum, is approximately equal to 1. This behavior is of particular interest when forming multiple emulsions, as detailed in Sections 2.2.3 and 3.4.3. Precisely, in the case of double emulsions, droplet formation is either a two-step process when $\{Ca, We\} < 1$ at both microchannel junctions (two dripping instabilities), or it is a one-step process when $\{Ca, We\} < 1$ at the first junction (dripping instability) while $\{Ca, We\} > 1$ at the second junction (jetting regime, no dripping instability), as detailed further in Section 2.2.4.

1.5.4 Dynamics of droplet interior and interfaces

While users new to the field of droplet microfluidics are intuitively aware that emulsion stability – basically the resistance against coalescence – and interfacial tension contributes to the overall capabilities of droplet microfluidics in materials design, analytics, and synthesis (cf. Section 3.5 for more on the matter of droplet stability), another property of emulsion droplets usually remains out of focus: their interface as well as interior is strongly dynamic [72]. That means, surfactant molecules constantly adsorb, desorb, and exchange with the surrounding droplet environment, while – above the CMC – surfactant micelles fuse with or form at the droplet interface owing to Ostwald ripening. In this process, solutes dissolved in the dispersed phase may partition into the continuous phase by direct transfer, being transported between droplets and into the continuous phase by surfactant micelles, or diffuse into neighboring droplets through a shared, porous surfactant bilayer [73].

Beyond its influence on interfacial tension and dynamics, the adsorption of surfactant molecules at the droplet's liquid–liquid boundary also leads to local rigidification and loss in surfactant mobility, respectively – a result of so-called

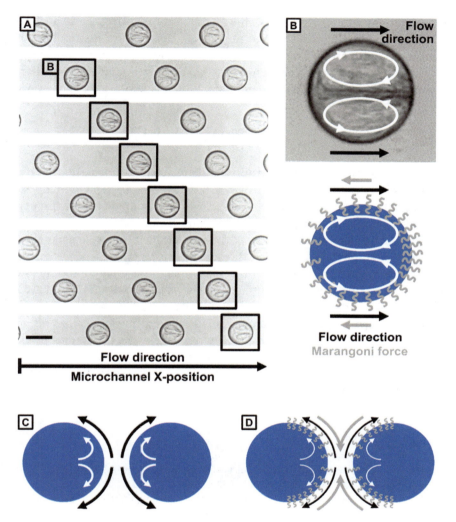

Figure 1.28: Flow pattern inside microdroplets and interfacial surfactant distribution during flow through a microchannel and after collection in a basin. (A) Fast camera image sequence of oil-in-water (O/W) emulsion droplets of 4-cyano-4'-pentylbiphenyl (5CB) in water with sodium dodecyl sulfate (SDS, 0.1% w/w) as a surfactant. The scale bar denotes 25 μm. (B) Single 5CB droplet flowing without constraints by surrounding microchannel walls (top micrograph), along with a schematic of the nonuniform surfactant distribution (gray) and internal fluid mixing pattern (lower sketch). Black arrows direct the flow direction, whereas gray arrows indicate a stress opposed to the flow, the so-called Marangoni force. (C) Without surfactants stabilizing their interface, droplets easily coalesce upon drainage of the continuous phase in between them. (D) Surfactant-stabilized emulsion droplets exhibiting a Marangoni stress (gray arrows), which counteracts removal of the continuous phase, thus improving the emulsion stability.

Marangoni convection [74, 75]. This effect arises from a difference in surface tension along the droplet surface. The underlying nonuniform distribution of surfactant molecules can be caused by fluctuations in temperature, concentration, or charge density. If the convective flux arises from a temperature difference, we can write the Marangoni number, Ma, which is another dimensionless quantity:

$$Ma = -\frac{\partial \gamma_{IF}}{\partial T}\frac{l\Delta T}{\eta a} \tag{1.70}$$

with the interfacial tension γ_{IF}, temperature T, the characteristic length l, the dynamic viscosity η, and the thermal diffusivity α (unit: $m^2 \; s^{-1}$). In a microfluidic experiment, we can observe Marangoni convection when microdroplets flow through a microfluidic device in a surrounding liquid (Figure 1.28A), for instance. The surfactant molecules accumulate at the front of the droplet, thereby locally generating a surface with low interfacial tension. Thereupon, the Marangoni effect causes a stress opposite to the flow direction at the rear of the droplet, with a locally low surfactant concentration and high surface tension, respectively (Figure 1.27B). While these emulsion droplets remain intact as long as they flow separate from each other, additional surfactants are required upon emulsion collection in a reservoir vessel. Only then, coalescence due to drainage of the continuous phase between two neighboring droplets is prevented (Figure 1.28C and 1.28D). Here, Marangoni convection contributes to emulsion stability by increasing the film drainage time between neighboring droplets, thus prolonging the emulsion lifetime.

References

[1] J. Kestin, M. Sokolov, W. A. Wakeham, "Viscosity of Water in the Range −8 °C to 150 °C", *J. Phys. Chem. Ref. Data* 1978, *7*, 941. **DOI: 10.1063/1.555581**.

[2] Y. M. Chen, A. J. Pearlstein, "Viscosity-temperature correlation for glycerol-water solutions", *Ind. Eng. Chem. Res.* 1987, *26*, 1670. **DOI: 10.1021/ie00068a030**.

[3] R. A. Clará, A. C. Gómez Marigliano, D. Morales, H. N. Sólimo, "Density, Viscosity, Vapor–Liquid Equilibrium, and Excess Molar Enthalpy of [Chloroform + Methyl tert-Butyl Ether]", *J. Chem. Eng. Data* 2010, *55*, 5862. **DOI: 10.1021/je100821g**.

[4] F. J. V. Santos, C. A. Nieto de Castro, J. H. Dymond, N. K. Dalaouti, M. J. Assael, A. Nagashima, "Standard Reference Data for the Viscosity of Toluene", *J. Phys. Chem. Ref. Data* 2006, *35*, 1. **DOI: 10.1063/1.1928233**.

[5] R. C. Hardy, "Viscosity of n-Hexadecane", *J. Res. Nat. Bureau Standards* 1958, *61*, 433. **DOI: 10.6028/jres.061.036**.

[6] M. J. P. Comuñas, X. Paredes, F. M. Gaciño, J. Fernández, J. P. Bazile, C. Boned, J. L. Daridon, G. Galliero, J. Pauly, K. R. Harris, M. J. Assael, S. K. Mylona, "Reference Correlation of the Viscosity of Squalane from 273 to 373 K at 0.1 MPa", *J. Phys. Chem. Ref. Data* 2013, *42*, 033101. **DOI: 10.1063/1.4812573**.

[7]   T. M. Aminabhavi, G. BinduDensities, "Viscosities, Refractive Indices, and Speeds of Sound of the Binary Mixtures of Bis(2-methoxyethyl) Ether with Nonane, Decane, Dodecane, Tetradecane, and Hexadecane at 298.15, 308.15, and 318.15 K", *J. Chem. Eng. Data*, 1994, *39*, 529. **DOI: 10.1021/je00015a029.**

[8]   S. N. Glasstone, K. Laidler, H. Eyring, "The Theory of Rate Processes", McGraw-Hill Book Company, Inc., New York (1941).

[9]   H. D. Weimann, "On the hole theory of viscosity, compressibility, and expansivity of liquids", *Kolloid Z. Z. Polym.* 1962, *181*, 131. **DOI: 10.1007/BF01499664.**

[10] A. K. Doolittle, "Studies in Newtonian Flow. II. The Dependence of the Viscosity of Liquids on Free-Space", *J. Appl. Phys.* 1951, *22*, 1471. **DOI: 10.1063/1.1699894.**

[11] M. H. Cohen, D. Turnbull, "Molecular Transport in Liquids and Glasses", *J. Chem. Phys.* 1959, *31*, 1164. **DOI: 10.1063/1.1730566.**

[12] P. B. Macedo, T. A. Litovitz, "On the Relative Roles of Free Volume and Activation Energy in the Viscosity of Liquids", *J. Chem. Phys.* 1965, *42*, 245. **DOI: 10.1063/1.1695683.**

[13] Richard A. L. Jones, Soft Condensed Matter, Oxford University Press (2002).

[14] C. V. Raman, "A Theory of the Viscosity of Liquids", *Nature* 1923, *111*, 532. **DOI: 10.1038/111532b0.**

[15] E. N. da C. Andrade, "XLI. A theory of the viscosity of liquids.–Part I", *Phil. Mag. S. 7*, 1934, *17*, 497. **DOI: 10.1080/14786443409462409.**

[16] E. N. da C. Andrade, "LVIII. A theory of the viscosity of liquids.—Part II", *Phil. Mag. S. 7*, 1934, *17*, 698. **DOI: 10.1080/14786443409462427.**

[17] R. H. Ewell, H. Eyring, "Theory of the Viscosity of Liquids as a Function of Temperature and Pressure", *J. Chem. Phys.* 1937, *5*, 726. **DOI: 10.1063/1.1750108.**

[18] G. Taylor, "Dispersion of Soluble Matter in Solvent Flowing Slowly through a Tube", *Proc. R. Soc. London, Ser. A* 1953, *219*, 186. **DOI: 10.1098/rspa.1953.0139.**

[19] G. Taylor, "Conditions under Which Dispersion of a Solute in a Stream of Solvent can be Used to Measure Molecular Diffusion", *Proc. R. Soc. London, Ser. A* 1954, *225*, 473. **DOI: 10.1098/rspa.1954.0216.**

[20] G. Taylor, "Diffusion and Mass Transport in Tubes", *Proc. Phys. Soc. B* 1954, *67*, 857. **DOI: 10.1088/0370-1301/67/12/301.**

[21] P.-G. de Gennes, "Soft Matter (Nobel Lecture)", *Angew. Chem. Int. Ed.* 1992, *31*, 842. **DOI: 10.1002/anie.199208421.**

[22] S. Seiffert, D. A. Weitz, "Controlled Fabrication of Polymer Microgels by Polymer-Analogous Gelation in Droplet Microfluidics", *Soft Matter* 2010, *6*, 3184. **DOI: 10.1039/C0SM00071J.**

[23] R. J. Poole, "The Deborah and Weissenberg numbers", The British Society of Rheology, *Rheology Bulletin* 2012, *53*, 32.

[24] M. Rubinbstein, R. A. Colby, "Polymer Physics", Oxford University Press (2003).

[25] Q. Ying, B. Chu, "Overlap concentration of macromolecules in solution", *Macromolecules* 1987, *20*, 362. **DOI: 10.1021/ma00168a023.**

[26] W. Burchard, "Solution Properties of Branched Macromolecules", *Adv. Polym. Sci.* 1999, *143*, 113. **DOI: 10.1007/3-540-49780-3_3.**

[27] B. H. Zimm, "Dynamics of Polymer Molecules in Dilute Solution: Viscoelasticity, Flow Birefringence and Dielectric Loss", *J. Chem. Phys.* 1956, *24*, 269. **DOI: 10.1063/1.1742462.**

[28] J. Brandrup, Edmund H. Immergut, E. A. Grulke (Edts), "Polymer Handbook", John Wiley & Sons, New York (1999).

[29] P.-G. De Gennes, "Dynamics of Entangled Polymer Solutions. I. The Rouse Model", *Macromolecules* 1976, *9*, 587. **DOI: 10.1021/ma60052a011.**

[30] S. Seiffert, "Functional Microgels Tailored by Droplet-Based Microfluidics", *Macromol. Rapid Commun.* 2011, *32*, 1600. **DOI: 10.1002/marc.201100342.**

[31] J. Thiele, "Polymer Material Design by Microfluidics Inspired by Cell Biology and Cell-Free Biotechnology", *Macromol. Chem. Phys.* 2017, *218*, 1600429. **DOI: 10.1002/macp.201600429**.
[32] P.-G. de Gennes, "Reptation of a Polymer Chain in the Presence of Fixed Obstacles", *J. Chem. Phys.* 1971, *55*, 572. **DOI: 10.1063/1.1675789**.
[33] R. W. Rendell, K. L. Ngai, G. B. McKenna, "Molecular weight and concentration dependences of the terminal relaxation time and viscosity of entangled polymer solutions", *Macromolecules* 1987, *20*, 2250. **DOI: 10.1021/ma00175a033**.
[34] A. R. Abate, M. Kutsovsky, S. Seiffert, M. Windbergs, L. F. V. Pinto, A. Rotem, A. S. Utada, D. A. Weitz, "Synthesis of Monodisperse Microparticles from Non-Newtonian Polymer Solutions with Microfluidic Devices", *Adv. Mater.* 2011, *23*, 1757. **DOI: 10.1002/adma.201004275**.
[35] Z. Yan, I. C. Clark, A. R. Abate, "Rapid Encapsulation of Cell and Polymer Solutions with Bubble-Triggered Droplet Generation", *Macromol. Chem. Phys.* 2017, *218*, 1600297. **DOI: 10.1002/macp.201600297**.
[36] A. Einstein, "Eine neue Bestimmung der Moleküldimensionen", *Ann. Phys.* 1906, *324*, 289. **DOI: 10.1002/andp.19063240204**.
[37] A. Einstein, "Berichtigung zu meiner Arbeit: „Eine neue Bestimmung der Moleküldimensionen", *Ann. Phys.* 1911, *339*, 591. **DOI: 10.1002/andp.19113390313**.
[38] E. Brown, H. M. Jaeger, "Shear thickening in concentrated suspensions: phenomenology, mechanisms, and relations to jamming", arXiv:1307.0269v1 [cond-mat.soft]. **DOI: 10.1088/0034-4885/77/4/046602**.
[39] N. J. Wagner, J. F. Brady, "Shear thickening in colloidal dispersions", *Physics Today* 2009, *62*, 27. **DOI: 10.1063/1.3248476**.
[40] J. Kaldasch, B. Senge, J. Laven, "Shear Thickening in Concentrated Soft Sphere Colloidal Suspensions: A Shear Induced Phase Transition", *Journal of Thermodynamics* 2015, 153854. **DOI: 10.1155/2015/153854**.
[41] B. Derjaguin, L. D. Landau, "Theory of the stability of strongly charged lyophobic sols and of the adhesion of strongly charged particles in solutions of electrolytes", *Prog. Surf. Sci.* 1993, *43*, 30. **DOI: 10.1016/0079-6816(93)90013-L**.
[42] E. J. W. Verwey, J. T. G. Overbeek, K. van Nes, "Theory of Stability of Lyophobic Colloids: The Interaction of Sol Particles Having an Electric Double Layer", Elsevier, New York (1948).
[43] W. H. Boersma, J. Laven, H. N. Stein, "Shear thickening (dilatancy) in concentrated dispersions", *AIChE Journal* 1990, *36*, 321. **DOI: 10.1002/aic.690360302**.
[44] O. K. Baskurt, H. J. Meiselman, "Blood rheology and hemodynamics", *Semin Thromb Hemost.* 2003, *29*, 435. **DOI: 10.1055/s-2003-44551**.
[45] A. M. Robertson, A. Sequeira, R. G. Owens, "Rheological models for blood", in: L. Formaggia, A. Qurteroni, A. Veneziani (Edts), "Cardiovascular Mathematics: Modeling and simulation if the circulatory system", Springer, Milan (2009).
[46] R. Brown: "A brief account of microscopical observations made in the months of June, July and August, 1827, on the particles contained in the pollen of plants; and on the general existence of active molecules in organic and inorganic bodies", *Phil. Mag.* 1828, *4*, 161. **DOI: 10.1080/14786442808674769**.
[47] A. Einstein, "Über die von der molekularkinetischen Theorie der Wärme geforderte Bewegung von in ruhenden Flüssigkeiten suspendierten Teilchen", *Ann. Phys.* 1905, *322*, 549. **DOI: 10.1002/andp.19053220806**.
[48] M. Smoluchowski, "Zur kinetischen Theorie der Brownschen Molekularbewegung und der Suspensionen", *Ann. Phys.* 1906, *326*, 756. **DOI: 10.1002/andp.19063261405**.
[49] A. Einstein, "Elementare Theorie der Brownschen Bewegung" *Z. f. Elektrochemie* 1908, *14*, 235. **DOI: 10.1002/bbpc.19080141703**.

[50] A. Fick, Ueber Diffusion, *Ann. Phys.* 1855, *170*, 59. **DOI: 10.1002/andp.18551700105**.
[51] J. B. Fourier, "Théorie analytique de la chaleur", English translation by A. Freeman, Cambridge University Press (1878).
[52] J. Crank, "The Mathematics of Diffusion", Clarendon Press, Oxford (1975).
[53] R. J. Hill, W. Konigsberg, G. Guidotti, L. C. Craig, "The structure of human hemoglobin. I. The separation of the alpha and beta chains and their amino acid composition", *J. Biol. Chem.* 1962, *237*, 1549.
[54] H. P. Erickson, "Size and Shape of Protein Molecules at the Nanometer Level Determined by Sedimentation, Gel Filtration, and Electron Microscopy", *Biol. Proced. Online* 2009, *11*, 32. **DOI: 10.1007/s12575-009-9008-x**.
[55] J. P. Brody, P. Yager, R. E. Goldstein, "Biotechnology at low Reynolds numbers", *Biophys. J.* 1996, *71*, 3430. **DOI: 10.1016/S0006-3495(96)79538-3**.
[56] J. P. Brody, P. Yager, "Diffusion-based extraction in a microfabricated device", *Sens. Actuators A* 1997, *58*, 13. **DOI: 10.1016/S0924-4247(97)80219-1**.
[57] W. Ostwald, "Über die vermeintliche Isomerie des roten und gelben Quecksilberoxyds und die Oberflächenspannung fester Körper", *Zeitschr. Phys. Chem.* 1900, *34*, 495. **DOI: 10.1515/zpch-1900-3431**.
[58] T. Young, "An Essay on the Cohesion of Fluids", *Phil. Trans. R. Soc. Lond.* 1805, *95*, 65. **DOI: 10.1098/rstl.1805.0005**.
[59] R. Dreyfus, P. Tabeling, H. Willaime, "Ordered and discordered patterns in two phase flows in microchannels", *Phys. Rev. Lett.* 2003, *90*, 144505. **DOI: 10.1103/PhysRevLett.90.144505**.
[60] T. Tadros, "Gibbs Adsorption Isotherm" in Encyclopedia of Colloid and Interface Science, Springer, Berlin Heidelberg 2013, 626. **DOI: 10.1007/978-3-642-20665-8_97**.
[61] R. G. Laughlin, "The Aqueous Phase Behavior of Surfactants", Academic Press Ltd., London (1994).
[62] T. Thorsen, R. W. Roberts, F. H. Arnold and S. R. Quake, "Dynamic pattern formation in a vesicle-generating microfluidic device", *Phys. Rev. Lett.* 2001, *86*, 4163. **DOI: 10.1103/PhysRevLett.86.4163**.
[63] H. A. Stone, Dynamics of Drop Deformation and Breakup in Viscous, *Annu. Rev. Fluid Mech.* 1994, *26*, 65. **DOI: 10.1146/annurev.fl.26.010194.000433**.
[64] P. B. Umbanhowar, V. Prasad, D. A. Weitz, "Monodisperse Emulsion Generation via Drop Break Off in a Coflowing Stream", *Langmuir* 2000, *16*, 347. **DOI: 10.1021/la990101e**.
[65] C. Cramer, P. Fischer, E. J. Windhab, "Drop formation in a coflowing ambient fluid", *Chem. Eng. Sci.* 2004, *59*, 3045. **DOI: 10.1016/j.ces.2004.04.006**.
[66] A. S. Utada, E. Lorenceau, D. R. Link, P. D. Kaplan, H. A. Stone, D. A. Weitz, "Monodisperse double emulsions generated from a microcapillary device", *Science* 2005, *308*, 537. **DOI: 10.1126/science.1109164**.
[67] S. L. Anna, N. Bontoux, H. A. Stone, "Formation of dispersions using 'flow focusing' in microchannels", *Appl. Phys. Lett.* 2003, *82*, 364. **DOI: 10.1063/1.1537519**.
[68] A. S. Utada, A. Fernandez-Nieves, H. A. Stone, D. A. Weitz, "Dripping to jetting transitions in coflowing liquid streams", *Phys. Rev. Lett.* 2007, *99*, 094502. **DOI: 10.1103/PhysRevLett.99.094502**.
[69] A. S. Utada, A. Fernandez-Nieves, J. M. Gordillo and D. A. Weitz, "Absolute Instability of a Liquid Jet in a Coflowing Stream", *Phys. Rev. Lett.* 2008, *100*, 014502. **DOI: 10.1103/PhysRevLett.100.014502**.
[70] J. Plateau, "Statique experimentale et theorique des liquides soumis aux seules forces moleculaires". Gauthiers-Villars, Paris (1873).
[71] L. Rayleigh, "On the Instability of Jets", *Proc. London Math. Soc.* 1878, *s1–10*, 4. **DOI: 10.1112/plms/s1-10.1.4**.

[72] J.-C. Baret: "Surfactants in droplet-based microfluidics", *Lab Chip* 2012, *12*, 422. **DOI: 10.1039/c1lc20582j**.

[73] Y. Skhiri, P. Gruner, B. Semin, Q. Brosseau, D. Pekin, L. Mazutis, V. Goust, F. Kleinschmidt, A. El Harrak, J. B. Hutchison, E. Mayot, J.-F. Bartolo, A. D. Griffiths, V. Taly, J.-C. Baret: "Dynamics of molecular transport by surfactants in emulsions", *Soft Matter* 2012, *8*, 10618. **DOI: 10.1039/c2sm25934f**.

[74] A. Karbalaei, R. Kumar, H. J. Cho: "Thermocapillarity in Microfluidics – A Review", *Micromachines* 2016, *7*, 13. **DOI: 10.3390/mi7010013**.

[75] R. Seemann, J.-B. Fleury, C. C. Maass: "Self-propelled droplets", *Eur. Phys. J. Special Topics* 2016, *225*, 2227. **DOI: 10.1140/epjst/e2016-60061-7**.

2 Segmented-flow or droplet-based microfluidics

2.1 Basics

Chapter 1 of this book has introduced the physical fundament of microfluidics. In that context, we have seen that there is basically two variants of microfluidics: laminar co-flow and segmented flow. The latter variant provides a way to compartmentalize fluid portions with exquisite control and versatility; as such, this variant also provides exquisite prospect for applications in both analytics and synthesis, but it also comes along with a need for further theoretical consideration. All that is in the focus of this present Chapter 2, which focuses on segmented flow or **droplet-based microfluidics** with the aim to form designer emulsions.

Emulsions are mixtures of multiple (basically two) immiscible fluids, one dispersed within the other in the form of droplets. In general, to obtain such droplets, the immiscible fluids must be brought into contact and then be subjected to shear forces, commonly applied by mechanical agitation such as vigorous stirring or ultra-sonication. With these methods, however, the shear forces are not distributed homogeneously within the system, thereby resulting in a broad distribution of droplet sizes. Such a broad distribution is detrimental if these droplets shall serve for further use, for example, in view of quantity-controlled encapsulation of additive payloads within the drops or with regard to templating microparticles from them. This is because any sort of droplet-size polydispersity will cause further kinds of polydispersities in downstream products. As a result, it is desirable to achieve exquisite control on the droplet size and shape in the process of emulsification.

2.2 Drop-by-drop formation in microfluidics

2.2.1 Single emulsions

A method that allows emulsion droplets to be formed with exquisite control is segmented-flow (SF) microfluidics, often also named droplet-based microfluidics [1, 2]. The basic idea of this method has been outlined in Section 1.5.2. Two or more mutually immiscible fluids are flown through a system of microchannels wherein they meet at destined points of channel intersection. In the region near these points, the fluid flows exert shear onto each other, thereby becoming segmented, most typically in the form of droplets. In this process, other than through the application of global shear forces in bulk methods of emulsification, the *local* application of shear creates one drop at a time in a controlled fashion. As detailed in Section 1.5.2, this principle of droplet formation can be explained using a water faucet as an example. If we turn on a faucet at a low flow rate, water drips out one drop at a time. The drop size is a result of the balance between the surface forces of the pending drop and its weight, and therefore depends on the surface tension of the fluid and the size of the faucet. Since both these parameters are time constant, all drops dripping from a faucet exhibit a monodisperse size distribution.[1] The same principle is active in microfluidic channels that bring together multiple streams of immiscible fluids. Periodic breakup of an inner fluid stream – named **dispersed phase** – can be induced by **flow-focusing** with an outer immiscible carrier fluid – named **continuous phase**. When these two fluids meet, droplets, plugs, or slugs form due to a **balance of interfacial tension and the shear exerted by the outer phase**. This balance is influenced by just a few parameters, including the viscosities, polarities, and flow rates of the fluids as well as the dimensions of the microchannels. Again, as all these parameters are or can be adjusted to be time constant, microfluidic devices can produce droplets with monodisperse size distribution. Furthermore, as most of these parameters can be controlled by the experimentalist, **great control can be achieved on the droplet size in a range of tens to hundreds of micrometers**.

Which of the three types of flow patterns – droplets, plugs, or slugs – form inside a microfluidic system depends on the choice of pairs of immiscible fluids [3]. For droplet formation (Figure 2.1A), we require a dispersed phase that avoids wetting the microchannel walls, whereas the continuous phase should completely cover the flow cell's surface. For plugs (Figure 2.1B), both the dispersed and the continuous phase should not wet the microchannel walls to ensure the formation of plugs instead of droplets. Similar to droplets, still though, plugs are discrete fluid volumes. In contrast to that, we can move from segmented flow to a continuous train of interconnected fluid entities in the form of the third mode of device operation: slug formation (Figure 2.1C). This process requires fluid combinations where the dispersed phase exhibits a strong affinity for the microchannel walls (as it is the case, for example, for hexane, toluene, or fluorocarbon oil on naturally hydrophobic or on hydrophobically treated channel walls), whereas the carrier phase is composed of solvents or gases that do not wet the channel surface. As the resulting slugs are interconnected by a layer of the dispersed phase wetting the microchannel walls, the most important feature of segmented flow – the formation of individual and isolated compartments – is lost in this case.

---

[1] According to the National Institute of Standards and Technology (NIST), "a particle size distribution may be considered monodisperse if at least 90% of the distribution lies within 5% of the median size" (Particle Size Characterization, Special Publication 960–961, January 2001).

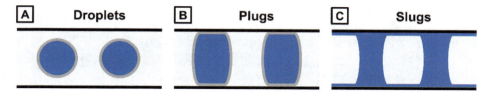

Figure 2.1: Three different types of flow patterns in microfluidic devices designated for emulsion formation. (A) Droplets separated from each other and isolated from the microchannel walls. (B) Droplets in contact with the surrounding channel walls, thereby plugging the cross-section of the microchannel, and hence, being referred to as plugs. Due to friction at the channel walls, the droplet interfaces appear convex. (C) Flow pattern of droplet-like slugs with concave shape that are in contact with neighboring slugs as well as with the microchannel walls. Adapted from [3]; copyright 2015 Royal Society of Chemistry.

Getting back to our faucet analogy: if we gradually increase the flow rate through the faucet, at a certain point, a thin water stream or a jet will be formed. This jet eventually also breaks up into drops due to a phenomenon named Rayleigh–Plateau instability [4], but these have a larger size range than those formed in the controlled dripping regime discussed above. Again, the same principle is operative in microfluidic channels, where dripping-to-jetting transitions occur at flow rates larger than certain critical ones, as elucidated in detail in Section 1.5.3.

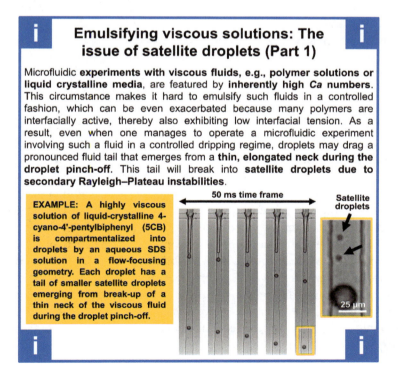

Emulsifying viscous solutions: The issue of satellite droplets (Part 1)

Microfluidic **experiments with viscous fluids, e.g., polymer solutions or liquid crystalline media**, are featured by **inherently high Ca numbers**. This circumstance makes it hard to emulsify such fluids in a controlled fashion, which can be even exacerbated because many polymers are interfacially active, thereby also exhibiting low interfacial tension. As a result, even when one manages to operate a microfluidic experiment involving such a fluid in a controlled dripping regime, droplets may drag a pronounced fluid tail that emerges from a **thin, elongated neck during the droplet pinch-off**. This tail will break into **satellite droplets due to secondary Rayleigh–Plateau instabilities**.

EXAMPLE: A highly viscous solution of liquid-crystalline 4-cyano-4'-pentylbiphenyl (5CB) is compartmentalized into droplets by an aqueous SDS solution in a flow-focusing geometry. Each droplet has a tail of smaller satellite droplets emerging from break-up of a thin neck of the viscous fluid during the droplet pinch-off.

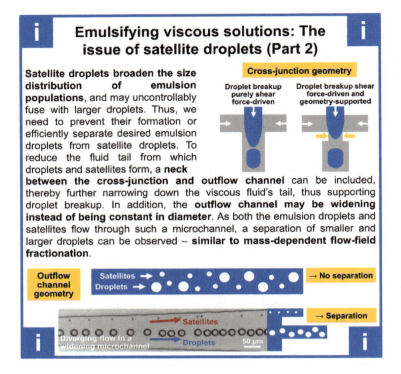

Intuitively, we envision the afore-described process of droplet formation in microfluidics as an *on-chip* process, where droplets form nonmiscible, leastwise pairs of fluids, for example, two liquids or a liquid and a gas, at flow-focusing T-, cross-, or co-flow microchannel junctions (Figure 2.2). Here, the ability to tailor the geometry as well as the surface properties of the microchannels on the same lengthscale as that of the droplets provides an accurately engineered reaction environment (Figure 2.2C). With the same degree of control over size and composition, microdroplets can likewise be formed *off-chip*, where a gas stream compartmentalizes liquids into a spray, for instance. Similar to droplet formation *on-chip*, spray properties are mainly determined by the flow-rate ratio of the continuous (gas) and the discontinuous (liquid) phase, the size of the spray-forming nozzle, as well as the microchannel wetting of the discontinuous phase inside the flow cell. Aside from the ability to generate droplets that are significantly smaller than in any current commercial spraying technique, common fabrication techniques of microfluidic devices also allow for precisely tailoring the spray nozzle design for individual liquids and solutes. Simply adapting and tailoring a typical microfluidic microchannel layout thus provides a significantly larger parameter space compared to the limited set of exchangeable nozzle designs that usually comes along with a commercial spray apparatus.

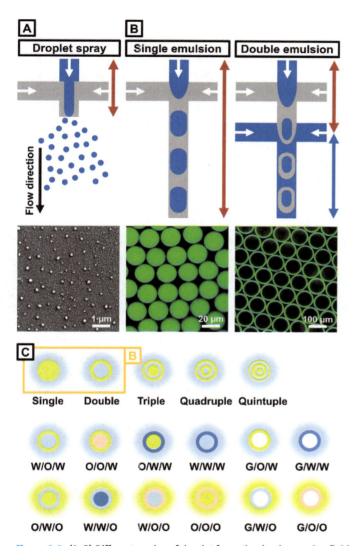

Figure 2.2: (A, B) Different modes of droplet formation in planar microfluidic devices (upper row), and examples of corresponding products (lower row). Blue and red arrows indicate hydrophilic and hydrophobic microchannel surface wettabilities, respectively, as required for forming emulsions with specific compositions. (A) Spray formation for preparing amorphous nanoparticles from hydrophobic, crystalline substances, such as drugs, by evaporative precipitation. (B) W/O single emulsion formation for encapsulating additive payloads, for example here, an artificial gene expression machinery that produces a green fluorescent protein (left row), as well as W/O/W double emulsion formation for generating amphiphile-laden droplet templates, which later transition into giant vesicles upon solvent evaporation (right row). (C) Structural variety in microfluidic emulsion formation, exemplarily showing all possible combinations of double-emulsion droplets. Adapted from [5]; copyright 2016 Royal Society of Chemistry.

2.2.2 Mixing phenomena in microdroplets

In Chapter 1, we have identified diffusion as the key mechanism for mixing across the liquid–liquid boundary in continuous-flow microfluidics. As the mixing time scales with the square of the distance over which diffusion occurs (→ Einstein–Smoluchoswki equation, see Section 1.3.5, eqs. 1.57 and 1.58), mixing is inherently slow in this situation. To reduce the mixing time, various **passive and active mixing geometries** have been developed that accelerate mixing by increase of the liquid–liquid contact area, for example, by induction of chaotic advection. In the latter case, mixing at low and intermediate Re numbers has been implemented by repeated folding and stretching of fluid streams. This treatment of fluids is based upon a baker's-like transformation, where dough is first lengthened by a factor of two and then folded over and over again; in microfluidic mixing, this process is readily repeatable until homogeneous mixing is achieved [6]. Similar to continuous fluid flow, we can employ the concept of folding, stretching, and reorienting fluids in microdroplets or plugs, which leads to an exponential decrease of the length of diffusive mixing. Although intuitively, mixing should be intrinsically fast within microdroplets, which are typically only a few picoliters in volume and a few tens of microns in diameter, respectively, there's still a plethora of reactions and processes that are so fast that they are completed before we achieve homogeneous mixing of all reaction partners in the droplets. As a result, insufficient mixing leads to heterogeneous material composition and functionalization, as exemplarily discussed for the preparation of polymer microgel particles from microdroplets in Section 3.5. To address this challenge, techniques to accelerate mixing within microfluidic droplets or plugs are useful and desirable.

To accelerate mixing in segmented flow, particularly by chaotic advection, we require fluid segments that are in contact with a microchannel wall – ideally fully filling the microchannel cross-section. Such a flow pattern will induce a steady recirculating flow inside the fluid plugs due to friction of the plugs' interfaces with the microchannel walls [7]. A different type of recirculating flow within microdroplets is observed if there is only insufficient contact between a droplet and the microchannel walls. For instance, if a microdroplet creeps along a microchannel that becomes gradually larger in cross-section size, as shown in Figure 2.3, we first observe a single circulating flow similar to a wheel rolling along a road. As the droplet detaches from the microchannel wall and freely floats within the microchannel volume, a second separate flow forms within the droplet, but flowing in inverse direction compared to the above example due to the absence of microchannel friction.

To promote mixing across the two halves of the recirculating flow inside the droplets, these should be passed through a meander-like microchannel design (see Figure 2.4) that promotes stretching, folding, and reorientation of fluid segments and thus resembles a baker's transformation [8]. Therefore, the choice of the microchannel design has important implications for materials design from droplet templates. For instance, insufficient mixing of prepolymer or monomer-containing microdroplets in

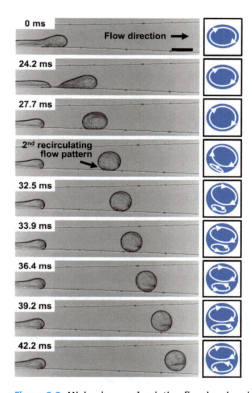

Figure 2.3: Mixing by a recirculating flow in microdroplets. The image sequence is a series of high-speed micrographs of a water-in-oil (W/O) droplet flowing through a microchannel (left column), along with corresponding schematics of the flow patterns inside the droplet (right column). The example displays the switching from a single recirculating flow extending through a droplet to two separate recirculating flows, each filling one-half of a droplet. If a droplet creeps along the microchannel wall – due to a wetting defect, for instance – it experiences strong asymmetrical friction at the surface inducing a single circular flow only. A second circular flow forms as the droplet detaches from the wall due to the changing flow field in the widening microchannel. Beyond the time frame of the image sequence, the droplet will adopt a symmetric internal flow field with two individual halves. The scale bar in the uppermost image denotes 50 µm and applies to all the other images as well.

a straight microchannel will yield Janus-like polymer particles with two distinguishable semispherical sides, for instance, whereas in a meander-type microchannel, efficient mixing will yield homogeneous particles from the same starting materials. For fast reacting species, mixing can be even further enhanced by utilizing meandering microchannels with different widths, as described in Section 3.5.

Droplet mixing is only one of many standard laboratory operations. To make microdroplets truly useful as independent, miniaturized reaction vessels, other standard operations for liquid manipulation need to be transferred from the macro- to the microscale. In view of this need, **microfluidic droplet-handling systems** for

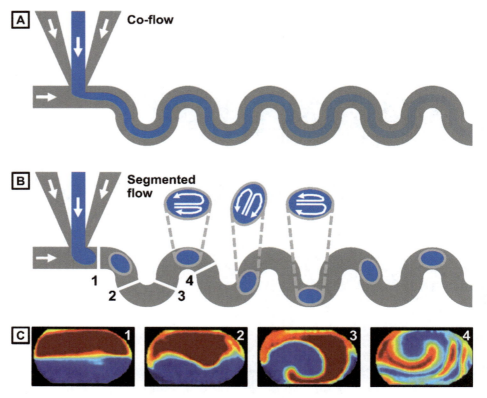

Figure 2.4: Mixing within a T-junction and a downstream meander-shaped microchannel. (A) Laminar flow of two miscible fluids. In this case, mixing occurs by diffusion across the liquid–liquid interface only. (B) Segmented flow of two immiscible fluids. In this case, mixing occurs in the individual droplets or plugs by chaotic advection. (C) Numerical simulation of chaotic mixing patterns inside surfactant-stabilized microdroplets at four exemplary positions in a meander-type microchannel, as indicated by white cross-sectional lines in Panel (B); adapted from [9]. Copyright 2012 American Institute of Physics Publishing.

incubating, sorting, and separating of the droplet microvessels have been introduced, allowing defined volumes of fluids to be injected and therefore mixed within preexisting droplets [10], as well as allowing droplets to be split [11, 12], merged [13, 14], incubated and re-injected [15], and to be sorted [16] at high rate and precision, as all depicted in Figure 2.5.

2.2.3 Higher-order emulsions

As we have discovered in Section 1.5.3 (practical details will be discussed in Section 3.4.2), oil-in-water (O/W) or water-in-oil (W/O) single emulsions can be

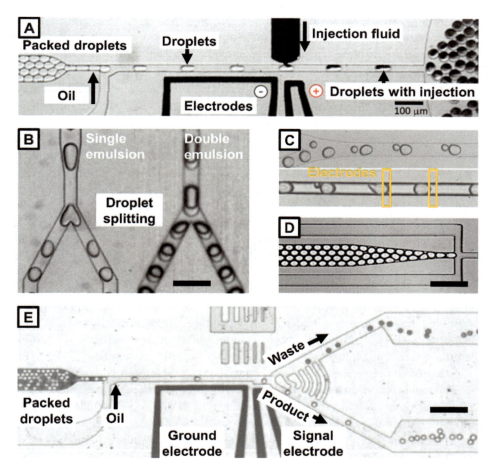

Figure 2.5: Microfluidic tools for droplet handling. (A) Injection of reagents into droplets by temporary destabilization of the droplet interface in a locally applied electric field. (B) Splitting of droplets into smaller daughter droplets by letting channel-spanning droplet entities hit a Y-junction. (C) Coalescence of droplets by local application of an electric field that destabilizes the droplet interface, similar to the principle of fluid injection into droplets shown in Panel A. (D) Droplet incubation within a basin channel. (E) Droplet sorting by on- and off-switching of an electric field. When switched on, this field directs the droplet into a product channel, whereas when switched off, the droplets move into a waste channel that has a lower resistance to the flow than the product channel. Reproduced from (A): [10], Copyright 2010 National Academy of Sciences; (B): [11], Copyright 2011 The Royal Society of Chemistry; (C): [13], Copyright 2006 American Institute of Physics; (D): [15], Copyright 2008 The Royal Society of Chemistry; (E): [16], Copyright 2010 National Academy of Sciences.

formed in a straightforward fashion (cf. Figure 1.26). However, single emulsions and materials that can be formed from such templates – notably polymer microparticles – are of limited use for encapsulation, protection, and controlled-release applications. This is because even though W/O and O/W emulsions can be tailored with respect to their size and shape, their stabilizing surfactant shell provides only limited control over encapsulation and release of entrapped additive payloads, and it is strongly dynamic (cf. Section 1.5.4). Moreover, polymer particles that are formed from single-emulsion templates are usually fabricated without any further substructure, and hence, material transport across the polymer particle boundary is difficult to control. Most typically, these polymer particles behave like continuous semipermeable entities, where diffusion of small molecules is merely controlled by the polymer particle porosity, whereas diffusive penetration of larger molecules (such as macromolecules) is blocked. Truly controlled encapsulation and release, however, is better achieved with a **core–shell capsule** with rationally tailored shell thickness and permeability, thereby allowing the uptake and release of additive payloads to be tailored as well. For this purpose, we require emulsions with an architecture that allows for templating such tailored capsules. These emulsions are higher-order or **multiple emulsions**, that is, emulsion droplets that are encapsulated inside other emulsion droplets, or in other words, "emulsions inside emulsions" [17].

While single-emulsion formation requires just one microfluidic "drop maker" (e.g., a flow-focusing junction in stamped flow cells), higher-order emulsions require a defined number of drop makers depending on the number of nested droplets that shall be formed. The most common higher-order emulsions are **double emulsions** of type oil-in-water-in-oil (O/W/O) or water-in-oil-in-water (W/O/W). For the sake of completeness, we need to add all-aqueous water-in-water-in-water (W/W/W) double emulsions to this list as well. The formation of such emulsions is possible if aqueous solutions of polymers are employed that naturally phase-separate from each other due to the very low polymer miscibility, thereby giving rise to aqueous two-phase systems (ATPS), for instance [18, 19]. In the following, however, we will focus on the more standard O/W/O and W/O/W emulsions.

Both stamped microflow cells and glass-capillary devices allow for multiple sequential or concurrent conduction of the process of emulsification, thereby forming higher-order emulsions with droplets containing subpopulations of smaller droplets inside, all with controllable monodisperse sizes and numbers. For example, when a glass microcapillary device consists of two tapered capillaries frontally facing one another in close proximity, three mutually immiscible fluids such as oil–water–oil or water–oil–water can be pumped through this arrangement, thereby forming drops-in-drops in a single step. In this situation, an inner fluid is pumped through a first, upstream cylindrical capillary with a tapered orifice, while a middle fluid, which is immiscible with the inner and outer fluids, flows into the same direction through an outer rectangular capillary that hosts and nests the injection tube, as shown in Figure 2.6. The outermost fluid flows through the same outer capillary but does so in the opposite direction from the other end, thereby hitting the coaxially flowing stream of the other two fluids at some point. This point of meeting and hitting of the fluid streams is at the tip of a second, downstream capillary that again has a cylindrical cross-section and

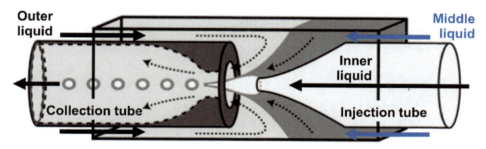

Figure 2.6: Formation of double emulsions in a glass-capillary microflow cell by the flow-focusing method. Adapted from [20] with permission from AAAS.

a larger tapered orifice. As a result, hydrodynamic flow-focusing of the two inner streams by the outer one occurs, thereby leading to formation of a double emulsion, which is a dispersion of monodisperse drops that each nest smaller monodisperse drops inside.

At low flow speed, the double-emulsion formation typically occurs approximately one tube diameter downstream from the entrance within the collection tube [20]. When drops form there, their size is controlled by the ratio of the flow rates of the combined inner and middle fluids to the outer fluid:

$$\frac{Q_{sum}}{Q_{of}} = \frac{R_{orifice}^2}{R_{orifice}^2 - R_{thread}^2} \qquad (2.1)$$

In eq. 2.1, Q_{sum} is the sum of the inner and middle fluid volumetric flow rates, Q_{of} is the flow rate of the outer fluid, R_{thread} is the radius of the fluid thread that breaks into drops, and $R_{orifice}$ is the radius of the collection tube where the drops are formed [20].

Double emulsions may also be formed in another class of microfluidic devices, those built from PDMS elastomers in a lithographic patterning process. In this variant, the devices feature two **cross-junction channels** in series. In the first junction, a single emulsion is formed. Downstream, in the second junction, this emulsion is subject to a second flow-focusing step, thereby forming a double emulsion with droplets inside droplets. Independent of whether double emulsions are formed in stamped flow cells or in capillary-based devices, for successful preparation, it is **crucial that the droplet formation between two drop makers – for example, flow-focusing junctions – is synchronized**. That means that a droplet formed at the first drop maker needs to arrive right in the moment of droplet pinch-off at the second drop maker to be successfully encapsulated by a liquid shell. We may achieve such synchronization of droplet formation by choosing a defined set of flow rates for the inner (core), the middle (shell), and the outer (continuous) phase, as exemplarily shown in Figure 2.7. The following Figure 2.8 summarizes the two key modes of droplet production, which are single-

and double-emulsion formation in stamped and in capillary-based microfluidic devices by the flow-focusing method.[2]

2.2.4 Modes of droplet formation: One-step emulsification versus stepwise emulsification

Higher-order or multiple emulsions (cf. Figure 2.2) are commonly formed in a sequence of flow-focusing junctions, where each additional junction adds another shell phase to the droplet core that is fabricated at the very first junction. For instance, **double emulsions made up of *two* phases and a continuous phase form in a series of *two* flow-focusing junctions**, as introduced in Section 2.2.3 (and exemplarily shown in Figure 1.23D). Conventionally, the formation of such multiple emulsions is a multistep process with a dripping instability at each droplet-forming junction. In this flow regime, monodisperse droplets form periodically and at fixed locations within the microfluidic device.

There are several **challenges in multiple-emulsion formation**. First, the **droplet formation needs to be synchronized** among all dripping instabilities to ensure a quantitative encapsulation efficiency of each dispersed-phase entity into its respective shell phase (cf. Figure 2.7). Second, the process-inherent need for alternating formation of O/W and W/O emulsions (or vice versa) in different areas of a microflow cell **requires the microchannels to be locally patterned** in view of their surface wettability.[3] Most of these patterning methods (some of these are presented in Section 3.4.1) are time-consuming and – particularly for new microfluidics users – rather difficult to implement with multiple sources of errors. To improve the availability of higher-order emulsions in microfluidics, it is thus necessary to simplify their formation. An obvious idea is to remove all dripping instabilities in the emulsion formation process, except for the very last one downstream. This way, all fluid phases would be encapsulated simultaneously independent of the flow conditions at all other microchannel junctions upstream. Indeed, by precisely tuning the flow rates in a microfluidic device, one can switch from a dripping-to-a-jetting

---

[2] Commonly, Newtonian fluids such as plain water and fluorinated oils are utilized for forming single and multiple emulsions due to the relative ease of droplet formation and to their strong visual contrast in microscopy, allowing for simple and contrast-rich observation of the microfluidic process by bright-field microscopy.

[3] One major exception is the formation of emulsions in nonplanar, that is, truly 3D microfluidic devices, which are designed such that they significantly widen in both width and height immediately downstream of the droplet pinch-off. If the flow rate of the continuous phase is sufficiently high in these flow cells, it can completely surround the innermost to-be-dispersed phase and prevent it from contacting the (far-away) microchannel walls. This way, droplet formation is feasible even if the dispersed phase preferentially wets the microchannel surface. The end of Section 2.3.3. provides further details on these special devices.

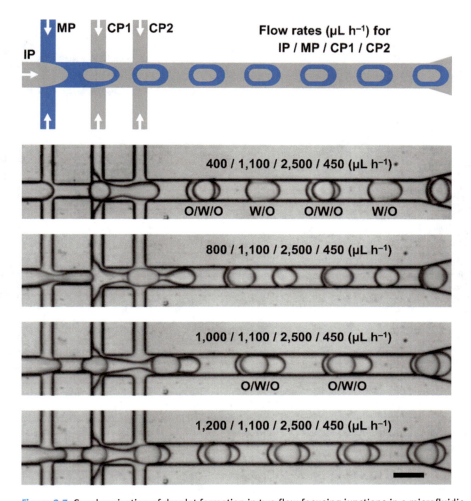

Figure 2.7: Synchronization of droplet formation in two flow-focusing junctions in a microfluidic device that produces an O/W/O double-emulsion by gradual optimization of the flow rates of three fluids, an inner phase (IP), a middle phase (MP), and an outer continuous phase (CP, composed of CP1 and CP2 in the example shown), from the uppermost to the lowermost micrograph. In the example shown, a third set of microchannels behind the two active drop makers allows for further fluid manipulation, for example, to realize an inflow of an additional continuous phase (CP2) in view to improve the double-emulsion pinch-off when non-Newtonian fluids (such as shear-thinning media) are used. This third set of channels can also allow for delayed addition of surfactants to speed up the emulsion formation, or it may serve to add further compounds such as polymerization accelerators that diffuse into the droplets formed upstream and then trigger their solidification by polymerization and gelation of monomers inside them. The images shown have been extracted from a high-speed imaging video recorded at 8,000 fps with 1.2×10^{-5} s exposure time of each frame. By tuning the inner-phase flow rate while keeping all other flow rates constant, droplet formation at the first and second flow-focusing junction is synchronized, such that quantitative droplet encapsulation is achieved. The scale bar in the lowermost image denotes 100 μm and applies to all the images above as well.

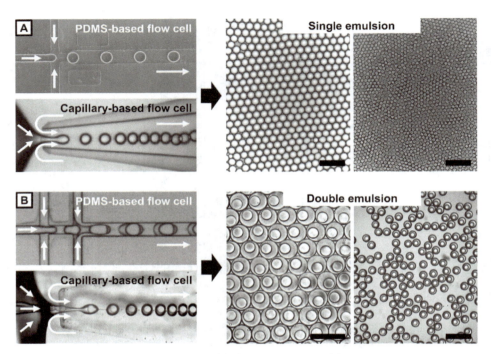

Figure 2.8: Microfluidic formation of monodisperse single- and double-emulsion droplets in elastomer-based microchannels and in glass microcapillary devices by flow-focusing. Arrows indicate the direction of flow. (A) An inner fluid phase is dispersed to breakup into monodisperse droplets by shear exerted by a second, immiscible outer fluid phase in a stamped PDMS-based microchannel device (upper images) and in a manually aligned glass microcapillary device (lower images). The scale bars in the right-hand images of (A) are 100 μm and 50 μm, respectively. (B) If the flow segmentation is carried out multiple times in a sequential or concurrent fashion, monodisperse double emulsions, which are droplets-in-droplets, are obtained, again in both a stamped PDMS-based microchannel device (upper images) and in a manually aligned glass microcapillary device (lower images). The scale bars in the right-hand images of (B) are 100 μm and 200 μm, respectively. Partly reprinted from [53]; copyright 2011 WILEY VCH.

regime at individual junctions, and thus remove dripping instabilities in multiple-emulsion formation in a controlled fashion (cf. Section 1.5.3). For instance, in double-emulsion formation with two flow-focusing junctions, we may set the flow rates to let the first junction jet, whereas the second junction is set to drip [21]. The key advantage of this so-called **one-step emulsion formation** is that surface modifications do not have to be as precise as in multistep emulsion formation. In double-emulsion formation, for instance, once the inner jet forms, it is surrounded by a protective sheath of the middle phase; this allows it to remain encapsulated even if the channel properties in that region favor wetting. This makes one-step formation easier to implement and, in general, more robust in practice.

To identify flow conditions that enable double-emulsion formation in a two-step process, and to delimit these from the flow conditions that enable double emulsion formation in a one-step process, we focus on two dimensionless quantities that have been introduced in Sections 1.5.2 and 1.5.3: the **Capillary number Ca** and the **Weber number We**. For the process of droplet break-off from a continuous liquid jet, We_{inner} describes the inertial force that pushes the inner-phase jet or the to-be-dispersed phase, respectively, forward in relation to its surface tension; Ca_{out} relates the shear force of a continuous-phase flow acting on the inner-phase jet to induce droplet-break-off, again in relation to its surface tension. By varying the inner-phase flow, and thus We_{inner}, while keeping the middle- and outer-phase flow constant, the occurrence of a dripping instability at an individual microchannel junction can be controlled. At slow inner-phase flow, and $\{We_{inner}, Ca_{outer}\} < 1$ in both microchannel junctions, a dripping instability can be observed in each junction (Figure 2.9, left) [22], causing emulsions to be formed in a two-step process. By contrast, as the inner flow rate is increased such that $We_{inner} > 1$, a coaxial jet of the inner and middle phase forms in the first junction [22], such that the inner phase is fully encapsulated by the middle phase, thereby minimizing the contact area with the microchannel walls (see above). As $\{We_{inner}, Ca_{outer}\}$ is still below one at the second junction, the continuous coaxial jet is segmented due to the remaining dripping instability. As a result, the inner and middle phases are encapsulated at the same time and place; therefore, double-emulsion formation becomes a one-step process.

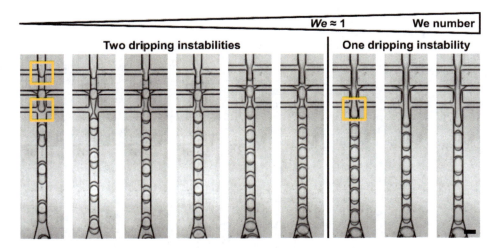

Figure 2.9: Double emulsion formation with two or one dripping instabilities, depending on the Weber number of the dispersed phase. The scale bar in the lower right denotes 100 μm and applies to all panels. Adapted from [21]; copyright 2011, Royal Society of Chemistry.

One-step emulsification of complex liquids (Part 1)

Removing dripping instabilities in multiple-emulsion formation does not only simplify their formation process and is experimentally less demanding, but it also **extends the range of application as well as the potential types of fluids from which double and higher-order emulsions can be made**. This feature is particularly important for microfluidics users who deal with complex liquids, which can be **viscoelastic like polymer solutions or which can exhibit a low interfacial tension like alcohols**. In the former case, polymers may evade droplet formation due to their elastic response under the shear exerted by the continuous phase at the microchannel junction. In the latter case, droplet formation is challenging because alcohols form a stable jet that usually extends far beyond a microchannel junction and is thus not easily emulsified.

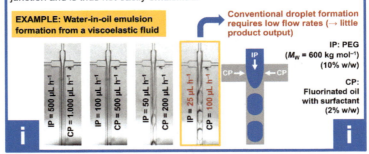

EXAMPLE: Water-in-oil emulsion formation from a viscoelastic fluid

Conventional droplet formation requires low flow rates (→ little product output)

IP: PEG (M_W = 600 kg mol^{-1}) (10% w/w)

CP: Fluorinated oil with surfactant (2% w/w)

One-step emulsification of complex liquids (Part 2)

Instead of attempting direct encapsulation of complex liquids under unfavorable experimental conditions, e.g., at extremely low flow rates or at a very high flow rate ratio of continuous and inner phase, it can be helpful to take an **experimental detour utilizing one-step emulsification**.

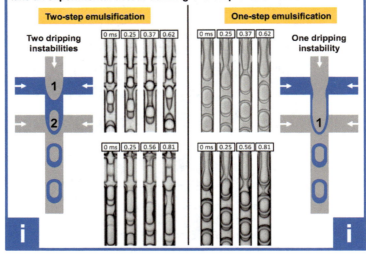

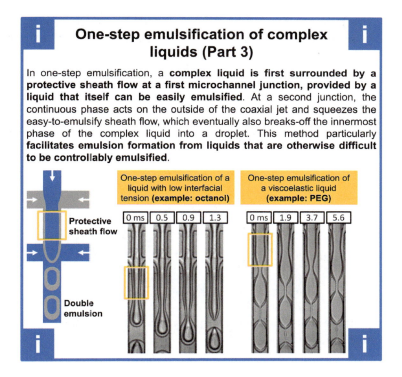

One-step emulsification of complex liquids (Part 3)

In one-step emulsification, a **complex liquid is first surrounded by a protective sheath flow at a first microchannel junction, provided by a liquid that itself can be easily emulsified**. At a second junction, the continuous phase acts on the outside of the coaxial jet and squeezes the easy-to-emulsify sheath flow, which eventually also breaks-off the innermost phase of the complex liquid into a droplet. This method particularly **facilitates emulsion formation from liquids that are otherwise difficult to be controllably emulsified**.

Just like for any multiple emulsion made by sequential emulsification, a key field of application of double emulsions made by one-step emulsification is their use as **templates** for forming **core–shell particles** (cf. Figure 2.19) as well as **capsules** in the area of drug delivery, protection of delicate goods, or multiphase materials. Here, the shell thickness of a core–shell particle or capsule strongly influences the material properties such as their release characteristics of encapsulated content, their shell degradability, or the capsule mechanics. It is thus of key interest to exert control over a particle shell thickness over a wide range and to enable tailoring of the shell characteristics to specific application needs.

To predict and tailor the shell thickness, $S_{thickness}$, of a double emulsion that may serve as a template for capsule or core–shell particle synthesis, we return to Figure 2.9, where double emulsion formation and its architecture is shown to be controlled by the Weber number of the inner phase, We_{inner}, and with that, by the inner-phase flow rate. The following paragraph focuses on deriving a link between We_{inner} and the shell thickness, thereby demonstrating that experimental observations (here, the decrease of the shell thickness with increasing We_{inner}) in microfluidics can be substantiated by simple mathematical equations. First, we identify all experimental parameters that influence double-emulsion formation on-chip. This list includes the *volumetric flow rates* of the inner, middle, and continuous phases, Q_{inner}, Q_{middle}, and Q_{cont}, and the *volumes* of the emulsion-droplets' inner and middle phases, V_{inner} and V_{middle}. We start with the parameter that is most closely related to the shell thickness, which is the middle-phase or shell volume of a double-emulsion droplet. V_{middle} can be described as the difference of the overall double-emulsion droplet and the inner-droplet volumes, which we set equal to the volumetric flow rate of the middle phase supplied over one drop-formation cycle within the time interval t ($[t]$ = s).

$$V_{\text{middle}} = V_{\text{double}} - V_{\text{inner}} = Q_{\text{middle}} \cdot t \tag{2.2}$$

As the droplet formation in the second microchannel junction is synchronized to the drop formation in the first junction, the time interval for one droplet formation can be described by the inner drop volume ($[V_{\text{inner}}]$ = m³) and the inner-phase volumetric flow rate ($[Q_{\text{inner}}]$ = m³ s⁻¹):

$$t = \frac{V_{\text{inner}}}{Q_{\text{inner}}} \tag{2.3}$$

With that, eq. 2.2 turns into:

$$V_{\text{middle}} = V_{\text{double}} - V_{\text{inner}} = \frac{Q_{\text{middle}} \cdot V_{\text{inner}}}{Q_{\text{inner}}} \tag{2.4}$$

Assuming that a double-emulsion droplet is a sphere with volume $V = \frac{4}{3}\pi r^3$, eq. 2.4 can be written as follows:

$$\frac{4}{3}\pi r_{\text{double}}^3 - \frac{4}{3}\pi r_{\text{inner}}^3 = \frac{Q_{\text{middle}} \cdot \frac{4}{3}\pi r_{\text{inner}}^3}{Q_{\text{inner}}} \Leftrightarrow r_{\text{double}}^3 = \frac{Q_{\text{middle}} \cdot r_{\text{inner}}^3}{Q_{\text{inner}}} + r_{\text{inner}}^3 \tag{2.5}$$

With $Q_{\text{inner}} = v_{\text{inner}} \cdot l^2$, eq. 2.5 reads:

$$r_{\text{double}}^3 = \frac{Q_{\text{middle}} \cdot r_{\text{inner}}^3}{v_{\text{inner}} \cdot l^2} + r_{\text{inner}}^3$$

The flow velocity of the inner phase, v_{inner}, can be written in terms of the Weber number of the inner phase:

$$We_{\text{inner}} = \frac{\rho \cdot v_{\text{inner}}^2 \cdot l}{\gamma_{\text{IF}}} \Leftrightarrow v_{\text{inner}} = \left(\frac{We_{\text{inner}} \cdot \gamma_{\text{IF}}}{\rho l}\right)^{\frac{1}{2}} \tag{2.6}$$

When inserting eq. 2.6 for Q_{inner} in eq. 2.5, we obtain:

$$r_{\text{double}}^3 = Q_{\text{middle}} \cdot r_{\text{inner}}^3 \left(\frac{We_{\text{inner}} \cdot \gamma_{\text{IF}} \cdot l^3}{\rho}\right)^{-\frac{1}{2}} + r_{\text{inner}}^3 \tag{2.7}$$

Taking the cube root gives:

$$r_{\text{double}} = \left(Q_{\text{middle}} \cdot r_{\text{inner}}^3 \left(\frac{We_{\text{inner}} \cdot \gamma_{\text{IF}} \cdot l^3}{\rho}\right)^{-\frac{1}{2}} + r_{\text{inner}}^3\right)^{\frac{1}{3}} \tag{2.8}$$

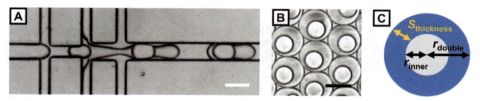

Figure 2.10: Formation of double-emulsion droplets and determination of their key geometric properties. (A) Double emulsions prepared in two sequential flow-focusing junctions via a two-step process (the third junction is not in use in the experiment shown; it may serve to add further components to the continuous phase for downstream droplet stabilization or modification). (B) As-prepared O/W/O double emulsion sitting on a glass-slide surface; the hexagonal packing of the droplets indicates their size monodispersity. (C) Shell thickness, $S_{thickness}$, corresponding to the difference of the double-emulsion's radius r_{double} and the radius of the inner droplet r_{inner}. The scale bars in Panels A and B denote 100 µm.

The middle-phase layer surrounding the inner droplet of the double emulsion has a shell thickness, $S_{thickness}$, as sketched in Figure 2.10.

$$S_{thickness} = r_{double} - r_{inner} \tag{2.9}$$

Inserting r_{inner} from eq. 2.8 into eq. 2.9 leads to:

$$S_{thickness} = \left(Q_{middle} \cdot r_{inner}^3 \left(\frac{We_{inner} \cdot \gamma_{IF} \cdot l^3}{\rho} \right)^{-\frac{1}{2}} + r_{inner}^3 \right)^{\frac{1}{3}} - r_{inner} \tag{2.10}$$

Factoring out r_{inner} gives

$$S_{thickness} = r_{inner} \left[\left(1 + Q_{middle} \cdot \left(\frac{We_{inner} \cdot \gamma_{IF} \cdot l^3}{\rho} \right)^{-\frac{1}{2}} \right)^{\frac{1}{3}} - 1 \right] \tag{2.11}$$

Equation 2.11 can be further simplified by defining a parameter a, which is a product of known constant parameters.

$$a = Q_{middle} \left(\frac{\rho}{\gamma_{IF} \cdot l^3} \right)^{\frac{1}{2}} \tag{2.12}$$

This leads to [21]:

$$S_{thickness} = r_{inner} \left[\left(1 + a We_{inner}^{-\frac{1}{2}} \right)^{\frac{1}{3}} - 1 \right] \tag{2.13}$$

In summary, the shell thickness only depends on We_{inner}, on known parameters, and on the inner-droplet radius, which is easily obtained from analysis of bright-field microscopy images, for instance.[4]

---

[4] For sufficient statistics, it may be necessary to determine the size of hundreds of droplets. For this purpose, software macros are available for ImageJ / Fiji or MATLAB.

> **Take-home messages Chapter 2.2**
>
> **Modes of droplet formation and manipulation**
>
> - Droplet microfluidics deals with the formation of liquid sprays, single emulsions, and multiple emulsions, e.g., droplets in droplets.
> - Multiple emulsions may be formed in a stepwise fashion or in a single step (depending on the *We* and *Ca* numbers); the latter process facilitates compartmentalizing complex liquids (e.g., fluids with low interfacial tension or non-Newtonian flow characteristics).
> - Although microdroplets are typically tiny (nano- to femtoliter volume) and mixing should be intuitively fast, additional geometrical features (e.g., meanders) may be required to achieve efficient homogenization of the droplet volume.
> - Emulsion droplets can be further manipulated by geometric features or electric fields to induce splitting, fusion, or separation from a larger droplet population.
> - Key applications utilize microfluidically prepared droplets for encapsulation and analysis down to single-molecule and single-cell levels, synthesis in microscopic reaction spaces, and as templates for capsules, microgels, and vesicles.

2.3 Experimental platforms for droplet formation

To realize microfluidic droplet formation, we may use different kinds of microchannel systems. The most popular ones are devices assembled from glass microcapillaries as well as devices lithographically molded from poly(dimethylsiloxane) (PDMS) elastomers.

2.3.1 Glass microcapillary assemblies

Capillary microfluidic devices consist of **coaxial assemblies of small glass tubes**, with typical lengths of a few centimeters and typical diameters in the submillimeter regime. One of the inherent advantages of these devices is that, since they are made from glass, their channel-wall wettability can be easily and precisely controlled by a surface reaction with an appropriate surface modifier. For example, a quick treatment with octadecyltrimethoxysilane will render the glass surface hydrophobic, whereas treatment with 2-[methoxy(polyethyleneoxy)propyl]trimethoxysilane will make the glass surface hydrophilic, as elucidated further in Section 3.1.7. An additional benefit of using glass is that the devices are both **chemically resistant and rigid**. To build these devices, a cylindrical glass capillary with an outer diameter of about 1 mm is heated and pulled to a tapered geometry that culminates in a fine orifice. The tapered capillary is then inserted into another glass capillary that has a square cross-section. Matching the inner and outer diameters of the square and the round capillaries provides coaxial alignment. Flowing one fluid inside the circular capillary while flowing a second fluid through the interstitial volume between then round and the square capillary in the same direction results in a three-dimensional coaxial flow of the two fluids (as illustrated in Figure 1.26A in Section 1.5.2). This variant is referred to as **co-flow geometry**. When both fluids flow at low rates, individual monodisperse drops are formed periodically at the tip of the capillary orifice, in a process termed dripping, as also shown in Figure 1.26A in Section 1.5.2. In the dripping regime, the drop at the end of the capillary tube experiences two competing forces: viscous drag pulling it downstream, and forces due to interfacial tension holding it to the capillary [22]. Initially, interfacial tension dominates, but as the attached droplet grows in size, drag forces eventually become comparable; this is when the droplet breaks off and is carried away by the flow of the continuous phase. Based on this picture, we may easily quantify the force balance at the point of droplet-breakup:

$$\eta_{out} \cdot v_{out} \cdot d_{droplet} \approx \gamma_{IF} \cdot d_{tip} \quad (2.14)$$

In eq. 2.14, v_{out} is the mean velocity of the outer fluid and γ_{IF} is the interfacial tension between the inner and outer fluid [22]. Based on this simple relation, we see that the diameter of the drop, d_{drop}, proportionally decreases as the outer-fluid flow rate increases. This interrelation holds until at a critical point the outer-fluid flow rate is so large that the first out of two possible dripping-to-jetting transitions detailed in Section 1.5.3 will occur.

An alternative geometry for drop formation in capillary devices is the **flow-focusing geometry**. In contrast to co-flow capillary devices, the two fluids are introduced from the two opposite ends of the same square capillary in the flow-focusing case (cf. Figure 1.26E). The inner fluid is hydrodynamically focused by the outer fluid through the narrow orifice of the tapered round capillary. At dripping conditions, drop formation occurs as soon as the inner fluid enters the circular orifice, whereas at jetting conditions, it occurs further downstream.

2.3.2 Poly(dimethylsiloxane) channel networks

A major drawback of glass microcapillary devices is that their manual fabrication makes it difficult to construct more than a few devices at a time. Moreover, the very nature of their fabrication by shaping the capillary tips by heating and stretching is limited in precision and reproducibility. The process therefore impairs producing large numbers of identical devices. A way to surpass these limitations is to use microfluidic devices molded from silicone elastomers in a process of **soft and photolithography** [23, 24], as detailed further in Sections 3.1.4–3.1.6. This method allows very small channels to be made with a level of precision that is superior to that of pulling glass capillaries; it also affords greater flexibility when designing devices with customized channel geometries. In short, the concept of this approach is to fabricate microfluidic devices from to-scale drawings that are printed on a transparency and then transferred to a photoactive resin. To mold a microfluidic device from the resulting positive relief, an elastomer such as poly(dimethylsiloxane) (PDMS) is poured over it and then hardened, peeled off, and bound to a solid carrier substrate. The result of this process is a device with micrometer-sized channels as drawn in the first step. As with glass microcapillary devices, a powerful way to produce drops in these channels is to use a flow-focusing geometry. This geometry has many variants, but all of them exhibit the same basic structure, in which two channels intersect to form a **cross-junction**. In this geometry, the dispersed phase is injected into the central inlet, whereas the continuous phase is injected into the two side inlets (cf. Figure 1.26F). Both fluids meet at the cross-junction where drops are formed. Again, the drop size is controlled by the fluid flow rates.

2.3.3 Dominance of wetting in PDMS devices

There is one main difference between glass microcapillary and PDMS channel-network devices: whereas droplet formation is mostly dominated by the coaxial three-dimensional channel geometry in the first type of devices, it is the channel-

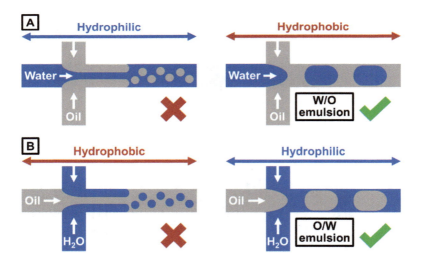

Figure 2.11: Influence of the microchannel wettability on the formation of emulsions in planar microfluidic devices. (A) Injection of water into a cross-junction with oil coming from both sides. The formation of water-in-oil (W/O) droplets, as shown in the right-side sketch, requires a hydrophobic microchannel surface to ensure complete wetting of the oil continuous phase, whereas a hydrophilic channel surface will form an opposite O/W emulsion in a way that the cross-junction operates like two independent T-junctions, as sketched in the left-hand schematic. (B) Opposite case of injection of oil and water into a cross-junction with hydrophobic surface. Again, the mismatched channel-wall wettability leads to a double-T-junction-like pinch-off of water droplets from both sides of the channel junction in the left-hand sketch, whereas if the surface wettability is hydrophobic, O/W emulsion droplets are formed with the cross-junction operating in a flow-focusing mode in the right-hand sketch.

wall **wettability** that determines the types of drops (oil-in-water versus water-in-oil) that are formed in the latter. The reason for this difference has to do with the way in which the fluids are injected into PDMS drop makers.

When the dispersed phase enters such a device, it is initially in contact with the surfaces of the channels. To be dispersed into droplets, it must be lifted off these surfaces and surrounded by the continuous phase. Whether this happens depends on the wettability of the channels. If, for example, the channels are hydrophobic, an inner water phase will be lifted off the walls by an outer oil phase, resulting in water drops (Figure 2.11A, right sketch). By contrast, if the channels are hydrophilic then an outer water phase can cling to the walls and lift off an inner oil phase, thereby producing oil drops (Figure 2.11B, right sketch). Wetting is of such importance in these devices that even if the inlets for the oil and water are switched, drops of the same type will be formed. For example, if water is injected into the outer-phase inlet and oil into the inner-phase inlet of a hydrophobic flow-focusing device, it will not be oil droplets but water droplets that will be formed, dripping from the two side channels, as though the device were two opposing T-junctions (Figure 2.11B, left sketch). Similarly, if oil is injected into the outer-phase inlet and water into the inner-phase inlet of a hydrophilic flow-focusing device, oil drops will be formed in that latter fashion (Figure 2.11A, left sketch). For these reasons, the wetting of PDMS devices must be controlled carefully when forming emulsions, and especially so when forming multiple emulsions. While **wettability control is rather simple to achieve for singe-emulsion formation** – here, we merely require the whole microfluidic device to be either completely hydrophilic (O/W emulsions) or hydrophobic (W/O emulsions) – **it is a major obstacle in double- and higher-order emulsion formation**.

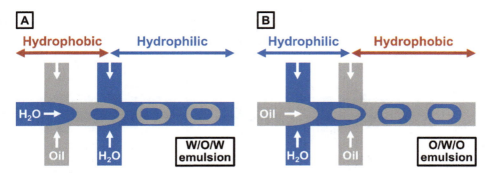

Figure 2.12: Schematic of double-emulsion formation with two dripping instabilities in planar microfluidic devices with spatially controlled microchannel surface wettability. (A) Water-in-oil-in-water (W/O/W) emulsions form in a sequence of two flow-focusing junctions, the first being hydrophobic and the second being hydrophilic. This way, water-in-oil single-emulsion droplets form at the first junction and are then encapsulated into a shell of water at the second junction, thereby yielding the desired double emulsions. (B) Similarly, oil-in-water-in-oil (O/W/O) emulsions form from two flow-focusing junctions, where the first is rendered hydrophilic and the second hydrophobic.

Surface modification of microflow cells towards tailored wettability (Part 1)

For mild reaction conditions that do not harm the integrity / shape of PDMS-based devices, **polyelectrolyte coatings** are sufficient. To render the microchannel wettability hydrophilic, liquid plugs of polyelectrolyte solutions with alternating charge like positively charged poly(allylamine hydrochloride) (PAH) and negatively charged poly(sodium 4-styrenesulfonate) (PSS) are flushed into defined areas of a flow cell. Over time, the polyelectrolytes pile up in a Layer-by-Layer (LbL) process forming a dense, hydrophilic coating.

While the LbL method is easy to apply, many applications of higher-order emulsions involve the use of organic solvents. These fluids potentially harm microflow cells by uncontrolled swelling of PDMS. On this account, **solvent-resistant, glass-like coatings** have been developed for PDMS microchannels that can be **additionally rendered hydrophilic by grafting hydrophilic polymers** like poly(acrylic acid) (PAA) onto their surface.

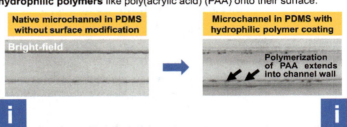

Surface modification of microflow cells towards tailored wettability (Part 2)

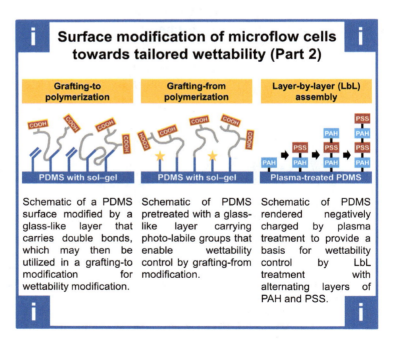

Schematic of a PDMS surface modified by a glass-like layer that carries double bonds, which may then be utilized in a grafting-to modification for wettability modification.

Schematic of PDMS pretreated with a glass-like layer carrying photo-labile groups that enable wettability control by grafting-from modification.

Schematic of PDMS rendered negatively charged by plasma treatment to provide a basis for wettability control by LbL treatment with alternating layers of PAH and PSS.

For instance, to form aqueous droplets in the first part of a device and then to encapsulate these into an oil shell in the second part of the same device – as required for W/O/W emulsion formation (Figure 2.12A) – microchannels need to be **locally modified with coatings** that provide micron-scale precision of the surface wettability and that can resist flow-induced shear forces and pressure over many hours. Such a modification of the microchannel wettability can be achieved by **surface polymerization** and **deposition of polymers** with hydrophilic groups (e.g., carboxyl groups, COOH) onto the microchannel surface or by layer-by-layer deposition of polyelectrolytes.

A more detailed view on such and other methods of tailoring of the surface properties of microfluidic devices, including methods utilizing plasma modification of the microchannels, will be provided in Section 3.4.1 [25–27]. Yet, independent of the afore-described approach, all examples require several additional fabrication steps in the prototyping of stamped flow cells, rendering microflow cell fabrication not quite rapid. In the future, fabrication of such rather complex flow cells may be simplified by the technique of **additive manufacturing**.

Despite the usual dominance of wetting in PDMS channel-network devices, there are strategies that can be used to relax these constraints. Using a **combination of geometrical control and shear**, it is possible to form either type of emulsion in a device with fixed wettability. These PDMS devices mimic the superior flow properties of glass-capillary devices by exploiting three-dimensional channel graduations. That way, they combine the best attributes of glass capillaries with the reproducibility and scalability of lithographically fabricated devices [28–30]. There is largely three realizations of such PDMS devices. The first one is a device with microchannels of multiple heights. For each channel junction, the inlet channel leading into the nozzle is smaller in its cross-sectional size than the nozzle, and it is centered vertically and horizontally with respect to the nozzle; this geometry largely mimics the one of coaxially aligned assemblies of glass microcapillaries. A simpler version of the device that works nearly as well but is easier to fabricate is to use a two-step, two-thickness approach. In this approach, the wetting of the device is of a uniform composition such that one type of emulsion is always preferred; for example, the device could be uniformly hydrophobic so that it easily forms water-in-oil emulsions. This is used in a first junction to form drops in a wettability-controlled first step, and then an abrupt channel widening is used in a second junction to form drops in a geometry-controlled second step, thereby resulting in drop-in-drop double emulsions. A third implementation is to use a one-step, two-thickness device. For example, all fluids can be injected into a single junction, where there is a single height step. This produces a coaxial jet of an inner and a middle phase, which is sheared into double emulsions by a sheath flow from the continuous phase. In yet a further variant of such geometry-controlled devices, use is made of a purely geometry-based pressure-drop instability that dictates a prescribed volume of incoming fluid to pinch-off as a droplet of predetermined size, which can be scaled-up to liters per hour.

> **Take-home messages Chapter 2.3 (Part 1)**
>
> **Experimental platforms for emulsion formation**
>
> - Conventional experimental platforms for microemulsion formation are largely based on glass capillaries or stamped microfluidic devices.
> - The advantage of glass capillary flow cells is their chemical inertness and their operability without need for sophisticated control of the microchannel surface wettability; their drawback is their manual fabrication that is limited to just one device at a time.
> - The advantage of stamped flow cells is their multiple replicative fabrication that can be parallelized easily; their drawback is the non-inertness and need for additional channel-surface wettability adjustment of the most common material used for device production, which is PDMS.

> **Take-home messages Chapter 2.3 (Part 2)**
>
> **Experimental platforms for emulsion formation**
>
> - In stamped, planar flow cells, higher-order emulsion formation requires spatially controlled wettability: W/O (O/W resp.) emulsions form in one part of the device, and are surrounded by an aqueous (oil, resp.) shell in another part.
> - Spatial control over the surface wettability with micrometer accuracy may be achieved by flow confinement, localized UV polymerization of hydrophilic polymers, or oxygen plasma treatment, as further detailed in Chapter 3.4.1.
> - Complex wettability patterns may be circumvented by utilizing geometry- and shear induced droplet breakup in non-planar flow cells mimicking the design of capillaries.

2.4 Application of segmented-flow microfluidics

2.4.1 Droplet-based microfluidics in analytics

The exquisite control on the formation of droplets, which are nothing else than sub-millimeter-sized fluid compartments, allows SF microfluidics to be used in great extent and versatility as a tool in the analytical sciences. This utility is further fueled by the rate of droplet formation in these systems, which can reach up to the kilohertz regime, thereby rendering droplet-based microfluidic systems useful as high-throughput analytical platforms. In these, reagents and analytes can be encapsulated into droplet compartments such to conduct chemical or biological assays, potentially semi- or fully automated. With this approach, especially **valuable and rare materials can be subject to analysis**, because when encapsulated in nanoliter-volume droplets, **even small amounts establish a high local concentration**. Furthermore, if subsequent analyses are conducted based on techniques that do not even require high concentrations to exhibit strong signals, for example, fluorescence-based methods, even trace amounts of substances down to the single-cell as well as single-molecule level can be analyzed. This is where SF microfluidic analytics unfolds its extreme strength. Based on that strength, microdroplet formation and compartmentalization have been combined with standard analytical methods. Beyond conventional fluorescence and confocal microscopy, SF microfluidics has been coupled with mass spectrometry analysis, where single-droplet content can be analyzed by on-chip extraction by electrocoalescence for ESI-MS or by single-droplet deposition on microarray plates for MALDI-MS, for example, for screening encapsulated enzymatic reactions [31–33].

Combination of these microfluidic tools (Figure 2.5) allows high-throughput screening to be realized on microfluidic handling platforms. Based on these methodological grounds, in a particularly spectacular branch of application, fully automated ultrafast sequencing of DNA [34] and RNA [35] was presented. For further enhancement, droplets can be first formed at such high dilution that only single or none of these valuable macromolecules are encapsulated within them, followed by amplification through in-droplet polymerase chain reactions [36–39] and sorting of desired populations, all automated such to optimize efficiency and rate. Similar protocols have also been used in various droplet-based microfluidic protein analyses [40]. In another branch of application, cell-sized droplets have been used as compartmentalizing environments to study stochasticity in gene expression at macromolecular crowding [41].

In all of the above examples, syntheses and analytics based on microemulsions proceed within droplets. Here, a surfactant shell protects the droplets from coalescence and prevents pronounced loss of encapsulated goods, which is why these are often discussed as individual reaction chambers. Nevertheless, several applications exist in which these reactions do not occur *inside* the droplet, but *on* its surface [42, 43]. Here, a biphasic reaction system of an organic continuous phase (referred to as "oil") and an aqueous dispersed phase can be utilized to increase the reactivity of organic solutes by stabilizing the transition state of a chemical reaction at the oil–water boundary by hydrogen bonding. While the surface-to-volume ratio and total interfacial area cannot be controlled in a reproducible fashion on macroscopic scales (e.g., in a stirred reactor), microfluidic W/O emulsions provide a defined interfacial area at which the adsorption of educts, a reaction, and eventually desorption of product can be studied based on the so-called on-water effect [44].

These are just a few examples for the extreme utility of SF or droplet-based microfluidics in analytics. There is many more, part of which reviewed in a series of overview articles on droplet-based microfluidic state of the art in general [45, 46], its fundamental physics [47], its use in single-cell [48, 49] and high-throughput analytics [50], and its further prospect in the fields of biology and medicine [51, 52].

2.4.2 Droplet-based microfluidics for particle design

A second broad class of application of droplet-based microfluidics is the use of droplets for the **design of advanced particulate materials**. The key idea of utilizing SF microfluidics for materials design is to **use the extremely monodisperse and uniform droplets** obtained by that method **as templates to form microparticles** in, on, or from them. These particles replicate the exceptional control over size, dispersity, shape, and internal organization of the microdroplet templates. Typically, microscopic particles are prepared from simple single-emulsion droplets (W/O or O/W) consisting of a pre-particle fluid – for example, a solution of low molar mass monomer molecules or macromolecular prepolymers – with typical sizes ranging from tens to hundreds of micrometers, along with size variations below 2%. Besides hard glassy or crystalline particles, soft microscopic hydrogels, which are called **microgels**[5] and consist of a hydrophilic polymer network swollen in water, are often templated from microfluidic emulsion droplets [53–55]. A selection of these and other particulate materials is shown in Figure 2.13. As microdroplets allow for short mixing times (owing to their nano- to femtoliter volume) and efficient heat transfer (due to their large surface-to-volume ratio), one can **achieve complete mixing of reaction components before nucleation, polymerization, and/or polymer crosslinking sets in**; on top of that, any reaction heat developing during these ways of droplet solidification can be efficiently dissipated. Both these characteristics are beneficial for the droplet solidification in a sense that they ensure well-defined conditions and therefore well-defined products; for example, in polymer microgel formation, the above conditions minimize polymer network heterogeneities and defects [56].

The solidification of droplets by means of monomer or oligomer polymerization can proceed via **thermally or enzymatically driven polymerization**. Yet, in the most prevalent case of microfluidic particle preparation, **UV-light** turns the droplets into particles by solidification of a compartmentalized reaction mixture containing

---

5 The original use of the term microgel referred to particles with sub-micrometer-scale colloidal dimensions, down to single, intramolecular crosslinked polymer coils (*Adv. Polym. Sci.* **1998**, *136*, 139). Later contributions have extended its use to larger, micrometer- or above-micrometer-scale particles (*Angew. Chem. Int. Ed.* **2007**, *46*, 1819). The IUPAC definition actually encompasses both these ranges (*Pure Appl. Chem.* **2007**, *79*, 1801) by defining the term microgel as a "particle of gel of any shape with an equivalent diameter of approximately 0.1 to 100 μm."

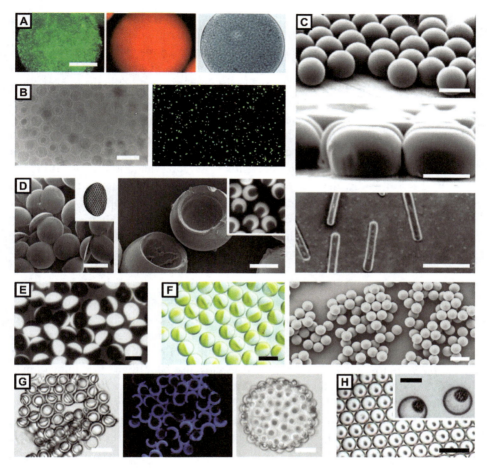

Figure 2.13: Polymer microparticles templated by single-emulsion droplet microfluidics. (A) Microgels with embedded additives such as fluorescent nanoparticles, quantum dots, and magnetic nanoparticles (from left to right). The scale bar denotes 100 μm and applies to all the three images. Reprinted from [57]; copyright 2007 WILEY VCH. (B) Fibroblast cells encapsulated inside microgels. A life–dead staining indicates high survival rates at 24 hrs after encapsulation. The scale bar for both panels is 100 μm. Reprinted from [58]; copyright 2014 RSC Publishing. (C) Spherical, pancake, and rod-shaped particles templated from droplets that are formed and solidified in channels that either fit them completely (top) or that squeeze them in one (middle) or two (bottom) directions perpendicular to that of the flow. The scale bars are 30 μm in the top and middle image and 400 μm in the bottom image. Reprinted from [59]; copyright 2005 WILEY VCH. (D) Nonspherical particles templated from anisotropic droplets solidified in one part only. Exemplarily, biphasic Janus droplets are made from a photo-polymerizable acrylate monomer (1,6-hexanediol diacrylate, HDDA) and conventional silicon oil, which can be selectively solidified by UV-irradiation (left), or from a so-called Aqueous Two-Phase System (ATPS), which spontaneously phase-separates to give rise to core–shell droplets. By employing a photo-polymerizable PEGDA and conventional dextran, photo-polymerization leads to selective solidification, and formation of microgels with an open socket (right). The scale bars denote 100 μm in the left panel and 30 μm in

a photoinitiator and a monomer right after the droplet formation on or off the microfluidic chip. Although, intuitively, UV-light should harm biological samples or cells that are encapsulated during the particle formation, mild reaction conditions can actually be provided by optimizing the UV-light intensity as well as the dwell time of the droplet templates during the UV exposure, thus broadening the range of applications of droplet-based particles for encapsulation (Figure 2.13A) [57, 58] toward cell biology (Figures 2.13B and 2.14) and cell-free biotechnology (Figure 2.14).

Beyond the formation of simple spherical particles, which represents the largest number of particle design applications in SF microfluidics, a **special strength of this technique is the control over complex flow patterns**, and thus its **ability of yielding complex droplet and particle morphologies**. For example, nonspherical morphologies are obtained by either squeezing the droplet templates during their solidification (for experimental details, cf. Section 3.4.2) [59] (Figure 2.13C) or by just partial solidification of the droplets. In the latter case, Janus-like droplets made from two different co-flowing liquids or double-emulsion droplets yield nonspherical microparticles if only one of the droplet constituents is solidified (Figure 2.13D) [60, 61]. With Janus-like droplets, whose two compartments can be both solidified, we may also form anisotropic particles with two distinct spheres, and thus very different material properties. For instance, these Janus-like particles may be both bicolored and electronically switchable (Figure 2.13E), with potential application as electronic ink, or they may exhibit anisotropic superparamagnetic properties (Figure 2.313F) [62, 63]. Anisotropic particles may also be formed by solidification of droplets that undergo a partial wetting transition due to solvent evaporation upon solidification (Figure 2.13G). Exemplarily, this technique allows for forming microgels that contain a hydrophilic and a hydrophobic compartment for utilization as particulate surfactants. In another approach, we may load single-emulsion droplets with inorganic nanoparticles that can be located in just one part of the droplet

Figure 2.13 (continued) the right panel. Left: reprinted from [60]; copyright 2007 WILEY VCH; right: reprinted from [61]; copyright 2012 WILEY VCH. (E, F) Janus particles templated by co-flow of two different but miscible pre-particle fluids followed by conjoint droplet formation in an external immiscible phase and then further followed by rapid droplet solidification. (E) Bicolored Janus microgels with one hemisphere containing carbon black, while the other hemisphere contains titanium oxide. The scale bar is 100 μm. Reprinted from [62]; copyright 2006 WILEY VCH. (F) Microgels made from co-flowing monomer solutions, of which one containing a ferrofluid for inducing magnetic anisotropy that allows for field-driven meshlike or chainlike microgel assembly. The scale bars denote 50 μm. Reprinted from [63]; copyright 2010 ACS Publications. (G) Anisotropic microgels made from single-emulsion droplets by solvent evaporation-induced partial wetting of fluids upon solidification. The scale bar is 10 μm for the left and middle image, and 20 μm for the right image. Reprinted from [64]; copyright 2010 ACS Publications. (H) Anisotropic microgels that can serve as micromotors formed by localization of patches of both magnetic and catalytically active nanoparticles by an external magnetic field upon off-chip solidification. The scale bar denotes 300 μm, and 50 μm for the inset. Reprinted from [65]; copyright 2016 ACS Publications.

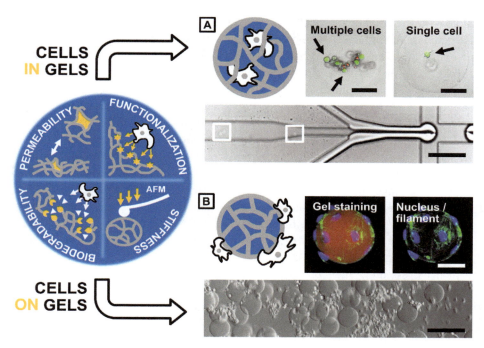

Figure 2.14: Microgels prepared from emulsion droplets allow for mimicking key features of the natural extracellular matrix (ECM) including stiffness, biodegradability (e.g., by cell secretion), permeability, and functionalization (e.g., by providing cell binding sites such as RGD) in a 3D environment. Key applications include studying growth, differentiation, migration, and communication of cells at low numbers or even at a single-cell level *inside* as well as cellular behavior depending on the substrate curvature *on* a microgel-based matrix. (A) Schematic of a polymer microgel as a 3D cell culture (left) (note that the polymer network meshes are not drawn to scale of the cell; in reality, they are about 5,000 times smaller than a cell), along with micrographs showing application of live–dead assays for staining human mesenchymal stem cells (hMSCs), which are cultured inside fibrinogen-functionalized microgel particles (right). Also shown is a bright-field microscopy image of co-encapsulation of a cell suspension with a prepolymer and crosslinker for subsequent microgel formation from droplet templates (bottom). The scale bar is 50 μm for the upper panels, whereas it is 100 μm for the bottom image. (B) Schematic of immobilized cells on a microgel surface (left), along with fluorescence micrographs of nucleus and cellular filament staining of cells attached to a microgel that is also fluorescently labeled (right). Also shown is a bright-field microscopy image of microgels functionalized with cell anchoring points that facilitate cell–microgel immobilization. The scale bar is 20 μm for the upper panels, whereas it is 100 μm for the bottom image.

template simply by gravity or external magnetic or electric fields to yield anisotropic microgels with a catalytically active compartment for utilization as self-propelled micro-swimmers in novel drug delivery concepts (Figure 2.13H) [64, 65].

The exact control over microgel formation from microdroplets with adjustable size, morphology, swelling properties, degradability, functionalization, and mechanical properties is of fundamental interest in many applications. Particularly in cell biology, microgels allow for cell immobilization or trapping of cell secretion by cell adhesion motifs or target-affine molecules that are incorporated in the microgel polymer backbone. Thus, such microgels serve as a **tailored artificial extracellular matrix (ECM)** that may expand our understanding of cellular functions from a two-dimensional (2D) environment on planar substrates toward a physiologically more relevant three-dimensional (3D) surrounding. Additionally, as cleavable groups can be easily included in the microgel architecture by adequate choice of the macromer from which the microgels are prepared, spatiotemporal control over degradation at a desired location is enabled, for example, for controlled release of encapsulated cells or drugs.

A particularly functional class of microgels are those that can react to external stimuli by change of their degree of swelling upon change of parameters such as pH, temperature, or salinity of the surrounding medium or by application of light or other external fields. This stimuli-sensitivity can be utilized to encapsulate – among others – biomolecules in the collapsed state of the microgel for subsequent release in the swollen state. In addition to that, in applications of microgels in cell biology, stimuli-responsive swelling and collapse of the microgel polymer network significantly changes the microgel elastic modulus, often simplified as stiffness. As cells are able to sense the stiffness of their surrounding (artificial, microgel-based) environment, such stimuli-responsive mechanical properties can convey important physical signals to cells through mechanotransduction pathways, influencing cell motility, proliferation, and differentiation. Overall, well-defined viscoelastic properties, spatiotemporal organization, permeability, and the presentation of cell-binding sites as well as labile network bonds that cells can potentially remodel in the process of growth, spreading, migration, and division are realized [66–68]. Some examples of microfluidically prepared hydrogel particles that reflect these properties in their role as advanced 3D cell culturing platforms are shown in Figure 2.14.

Hard and soft microparticles may also be obtained with other complex morphologies using sophisticated microfluidic methods of droplet templating. A particularly useful type of template is that of **core–shell double-emulsion droplets**, which intrinsically resemble the structure of **microcapsules** (Figure 2.15). A simple strategy of forming such capsule structures is to either solidify only the shell of a double-emulsion droplet while leaving the core unsolidified; in some cases, core–shell particles can even be obtained by solidifying the rim of single-emulsion droplets [69–73]. Other variants of core–shell particles are obtained by encapsulation of hydrogel particles serving as core material in a subsequent encapsulation step or by solidifying two immiscible liquids containing monomers or other crosslinkable species [73, 74].

The above examples of microgel particle and capsule preparation by droplet microfluidics are just a few, merely giving a glimpse on the extreme versatility and utility of microfluidics in particulate materials design. There is a lot more, some of which have been illustratively reviewed in a series of overview articles [76, 77].

Yet another strategy of forming microcapsules is to **build core–shell geometries by spontaneous or directed assembly of micro- or nanoscopic building blocks**

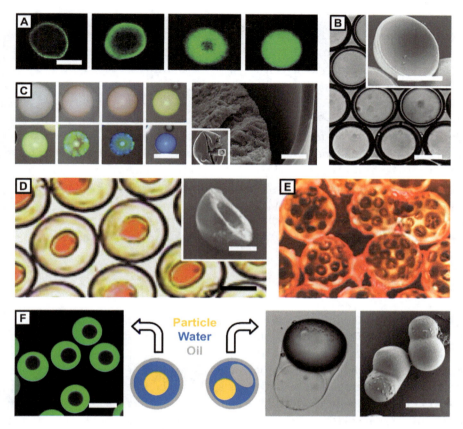

Figure 2.15: Capsules and core–shell particles prepared by droplet microfluidics. (A) Confocal fluorescence micrographs of alginate core–shell microgel capsules formed from single-emulsion drops that are gelled in their rim only by diffusive influx of Ca^{2+} ions from the external oil phase for different times. The scale bar for all panels denotes 20 μm. Reproduced from [69]; copyright 2006 American Chemical Society. (B) W/O/W double emulsions whose middle phase is prepared from a crosslinkable silicone monomer to yield inorganic capsules upon middle-phase solidification. The scale bar is 100 μm for all panels. Reproduced from [70]; copyright 2013 ACS Publications. (C) Photonic microcapsules prepared from W/O/W double emulsions containing colloidal particles made of polystyrene ($d \approx 175$ nm), whose shell phase is loaded with polymerizable trimethylolpropane ethoxylate triacrylate. Optical properties can be modified by the concentration of the colloidal particles in the aqueous core as well as the thickness and rigidity of the double-emulsion shell phase. The scale bar denotes 2 μm. Reproduced from [71]; copyright Macmillan Publishers. (D) Polymer capsules made from W/O/W double emulsions, whose shells are loaded with hydroxybutylacrylate and solidified by on-chip UV polymerization. The inner aqueous core is fluorescently labeled with a red dye to enhance the image contrast. The scale bar is 100 μm for the main image and 75 μm for the inset. Reproduced from [72]; copyright 2006 IOP Publishing. (E) Multi-core poly(lactid-*co*-glycolic acid) microgels loaded with the antibiotic gentamycin sulfhate. The scale is not specified. Reproduced from [73]; copyright 2005 WILEY VCH. (F) Core–shell microgels with one particle core (left) and two compartments (right). The scale bars denote 200 μm in the left part image and 25 μm in the right image. Left: reproduced from [74]; copyright 2010 American Chemical Society; left: reproduced from [75]; copyright 2011 Royal Society of Chemistry.

in or on templating microfluidic droplets. These processes lead to tailored formation of supraparticles such as colloidosomes [78], lipid vesicles (liposomes), and copolymer vesicles (polymersomes) [79–82], which consist of thin shells of colloidal nanoparticles or block copolymers (Figure 2.16). There is a **myriad of applications of vesicles prepared by droplet microfluidics:** they can be synthetic microreactors for confinement, manipulation, and optimization of (bio-)chemical reactions, they can serve as structural mimics for understanding and utilizing cellular functions, or they can find use as containers for protecting precious samples from degradation and enhancing their retention time in targeted (drug) delivery. All these applications have one thing in common: they all require vesicles with well-defined size, shell composition, compartmentalization, and high encapsulation efficiency. The following paragraphs focus on both the formation process of vesicles by droplet microfluidics and on experimental parameters that allow for tailoring the physicochemical and mechanical properties of the vesicles for the above-introduced key applications.

2.4.3 Droplet-based microfluidics for vesicle design

A powerful preparative area of application of multiphase co-flow microfluidics is vesicle formation at co-flowing liquid–liquid interfaces, as detailed in Section 3.2.2. This technique allows for reproducible polymersome or liposome preparation in the nanometer-range, but with limited encapsulation efficiency. Vesicle formation by droplet templates complements the size range of vesicles to be formed to approximately 20–200 µm, and – most importantly – allows for preparing vesicles with nearly **quantitative encapsulation efficiency**. This is because the vesicles are formed from W/O/W double-emulsion droplets, whose inner-to-middle and middle-to-outer interface is stabilized against coalescence by the same amphiphiles (lipids or copolymers) that also serve as the vesicle building blocks. As a result, removal of the oil phase separating the amphiphile single layers of the double emulsion by evaporation causes them to assemble into an amphiphile bilayer and vesicle membrane, respectively. In this process, the W/O/W emulsion's inner aqueous phase becomes the aqueous core of the as-formed vesicle. Thus, a vesicle membrane will fully encapsulate whatever has been injected as aqueous solution into the microfluidic device before (Figure 2.17A). Another key feature of emulsions is that they direct the assembly of amphiphiles at their O/W and W/O interfaces. Thus, we can utilize such amphiphiles that naturally do not form vesicles due to an unfavorable packing parameter P (further information on this parameter can be found in Section 3.2.2).

Yet, while amphiphile-stabilized double emulsions provide a straightforward way to ~100-µm sized vesicles with excellent encapsulation efficiency, their transition is delicate. Only if key parameters – solvent composition, osmotic pressure, amphiphile concentration, and interfacial tensions – and their interplay are understood, the desired transition from emulsions to vesicles can be observed.

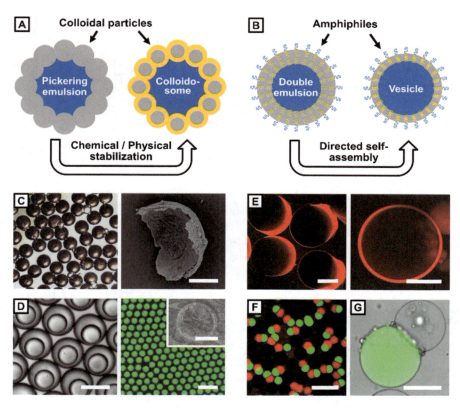

Figure 2.16: Formation of colloidosomes and vesicles from droplet templates. (A) Single-emulsion droplet whose W/O or O/W interface is stabilized by colloidal particles (named Pickering emulsion), serving as a template for colloidosome formation. By either covalent linking, thermal annealing of the particle shell, or addition of electrolytes to complex oppositely charged colloidal particles, the droplet transitions into a colloidosome. (B) W/O/W double-emulsion droplet whose inner-to-middle and middle-to-outer interface, respectively, is stabilized by amphiphiles, serving as a template to guide the assembly of the two single amphiphile layers into a vesicle bilayer. (C) Colloidosomes prepared from W/O emulsion droplets loaded with chitosan and silicon oxide nanoparticles. The scale bar is 100 μm in the right image. Reproduced from [83]; copyright 2016 Royal Society of Chemistry. (D) Colloidosomes made from W/O/W double emulsions stabilized by hydrophobic silica nanoparticles, which stabilize the emulsion templates against coalescence and serve as colloidal building blocks after solvent evaporation. The scale bar denotes 100 μm in the left image, and 500 μm for the right image; the scale bar for the inset is 50 μm. Reproduced from [78]; copyright 2008 WILEY VCH. (E) Polymersomes prepared from the block copolymer poly(ethylene glycol)-*block*-poly(lactic acid) (PEG-*b*-PLA) and fluorescently labeled with Nile Red. Excess amphiphiles that aggregate on the vesicle surface (left) may occasionally detach (cf. Section 3.4.5). The scale bars are 20 μm. Reproduced from [81]; copyright 2010 WILEY VCH. (F, G) Two-compartment polymersome prepared from double emulsions with two aqueous inner droplets. (F) Liposomes made from the lipid L-α-phosphatidylcholine; the two compartments are fluorescently labeled with Alexa Fluor 488 and 647, respectively. The scale bar is 200 μm. Reproduced from [84]; copyright 2016 ACS Publications. (G) PEG-*b*-PLA-polymersomes; one compartment is labeled with FITC dextran. The scale bar denotes 50 μm. Reproduced from [80]; copyright 2011 WILEY VCH.

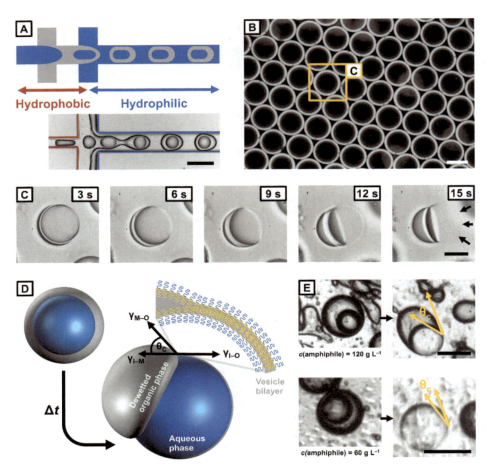

Figure 2.17: Formation of microscopic vesicles from amphiphile-stabilized double emulsions. (A) Microflow cell for forming W/O/W double emulsions with spatially controlled wettability. (B) Double emulsions stabilized by a block copolymer-type amphiphile. The hydrophobic middle phase is loaded with a fluorescent dye. Hexagonal packing of the emulsion droplets indicates their size monodispersity. The scale bar for (A) and (B) is 50 µm. (C) Image sequence of a double-emulsion droplet undergoing a wetting transition upon collection in a sealed microscope observation chamber. The scale bar denotes 50 µm for all panels. (D) Depending on the amphiphile concentration in the initial organic-solvent mixture, an equilibrium structure with a distinct contact angle at the three-phase contact point (inner-to-middle with inner-to-outer with middle-to-outer phase) forms. Typically, the equilibrium structure has an acorn-like shape composed of an organic-solvent droplet with excess amphiphile in contact with a solvated vesicle membrane, highlighted by arrows in (C). (E) Morphologies of single dewetted double-emulsion droplets with different contact angles at the tree-phase contact point for two distinct amphiphile concentrations. Generally, at higher amphiphile concentrations in the organic solvents, larger contact angles can be observed in the equilibrium structure of the dewetted double emulsion. The scale bar is 50 µm.

Along these lines, the following section compiles the scattered knowledge on the physical chemistry of double-emulsion-to-vesicle transition and aims at providing a detailed understanding of the different driving forces involved in this process.

Naturally, we start from amphiphile-stabilized W/O/W double-emulsion droplets (Figure 2.17B). After collection of these, the organic solvent phase and the double emulsion's shell will soon evaporate, promoted by the large surface-to-volume ratio of the emulsion droplets and the partial solubility of organic solvents in the surrounding aqueous phase (e.g., for chloroform, we have a solubility of 8 g L^{-1} in water at 20 °C). To account for the destabilizing effect of a density mismatch between the double-emulsion droplets and the surrounding continuous phase, and to tailor the evaporation rate of the double emulsion's middle phase during the process of emulsion-to-vesicle transition, the amphiphiles are commonly dissolved in a precisely balanced mixture of at least two organic solvents. As the evaporation of the solvent mixture progresses, a **dewetting transition** can be observed, as shown in Figure 2.17C. In this process, the organic-solvent phase dewets from the inner aqueous phase, leading to a partial wetting scenario. Its final equilibrium state – usually exhibiting an acorn-like shape – consists of a vesicle bilayer plasticized by remaining organic solvent, while excess amphiphile is trapped in an adherent droplet of organic solvent. In the acorn-like structure, we find a **three-phase contact point**, as indicated in Figure 2.17D, where the inner, middle, and continuous phases coexist. This three-phase contact can be expressed in terms of

$$-S_P = \gamma_{IF,I-M} + \gamma_{IF,M-O} - \gamma_{IF,I-O} \qquad (2.15)$$

Where S_P is named *spreading parameter*. If the organic-solvent phase wets the vesicle membrane after the dewetting transition only partially, S_P is negative. In contrast, at complete wetting, S_P is a positive number [85, 86].

The interplay of the interfacial energies during the dewetting transition also gives rise to an attraction, which is the adhesion energy $E_{adhesion}$ [87]:

$$E_{adhesion} = \gamma_{IF,I-M} + \gamma_{IF,M-O} - \gamma_{IF,I-O} \qquad (2.16)$$

with the interfacial tension of the bilayer $\gamma_{IF,I-O}$, the interfacial tension of the inner-to-middle and middle-to-outer interface, $\gamma_{IF,I-M}$ and $\gamma_{IF,M-O}$, respectively, and the adhesion energy, $E_{adhesion}$, that arises between the inner and outer aqueous phase of the double emulsion. Due to the similar composition at the inner-to-middle as well as the middle-to-outer interface, the interfacial energy is comparable, such that $\gamma_{IF,I-M} \approx \gamma_{IF,M-O}$ can be assumed. Thus, the contact angle θ between the W/O and O/W interfaces (cf. Figure 2.17D), and ultimately the morphology of the equilibrium structure in the state of partial wetting can be directly determined using the Young–Duprè equation.

$$E_{\text{adhesion}} = 2\gamma_{\text{IF, M–O}}(1 - \cos\theta) \tag{2.17}$$

The adhesion energy E_{adhesion} depends on the physicochemical properties of the dissolved amphiphile and can be expressed by

$$E_{\text{adhesion}} = a\xi\pi_{\text{osm}} \tag{2.18}$$

with

$$\pi_{\text{osm}} = ck_B T \text{ (for dilute amphiphile solutions)} \tag{2.19}$$

Here, π_{osm} is the osmotic pressure,[6] and a is a numerical coefficient that depends on the amphiphile concentration regime. ξ is the size scale of the amphiphile. For instance, in dilute polymer solutions, $a = \frac{4}{\sqrt{\pi}}$, and ξ corresponds to the radius of gyration R_g. In this case, E_{adhesion} scales linearly with the amphiphile concentration c (in contrast to semidilute solutions, where $E_{\text{adhesion}} \sim c^{3/2}$) [87].

The influence of the amphiphile concentration on double emulsion-based vesicle formation (Part 1)

The **amphiphile concentration is a crucial parameter in the process of vesicle formation** for three reasons:

1 A minimal amphiphile concentration is required to fully occupy the W/O ad O/W interface to resist droplet coalescence.

The minimal amount of amphiphiles depends on its volume occupancy at the liquid–liquid interface. Experiments show that excess amphiphile is beneficial to ensure emulsion and vesicle stability under shear, for instance, and to adjust the bilayer thickness as well as the vesicle structure.

2 The amphiphile concentration provides additional control over the morphology of the final vesicles.

While there is an optimal window of amphiphile concentration for vesicle formation, further increase of the amphiphile concentration for forming double emulsions can provide access to other morphologies and structures, ranging from vesicles to amphiphile capsules.

---

6 The osmotic pressure π_{osm} of the inner and outer double-emulsion phase must be balanced to advert osmosis-induced instability of the droplet template – which would exacerbate vesicle size dispersity – and shrinkage / collapse of the later vesicles.

The influence of the amphiphile concentration on double emulsion-based vesicle formation (Part 2)

3 The actual driving force for the dewetting transition from a double emulsion to a vesicle are depletion interactions.

These **depletion interactions arise between the amphiphile-stabilized W/O and O/W interfaces of the double emulsion**. The magnitude of these forces scales with the amphiphile concentration c, and likewise determines the morphology of the partial wetting state via the contact angle θ following the Young–Duprè equation. The **origin of the depletion forces lies in dissolved amphiphile molecules within the thin organic solvent layer**, which are driven out into the as-formed organic solvent droplet because the excluded volume surrounding the amphiphile-stabilized W/O and O/W interfaces overlap.

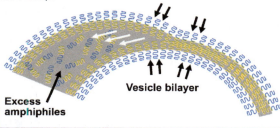

Excess amphiphiles / Vesicle bilayer

The influence of the amphiphile concentration on double emulsion-based vesicle formation (Part 3)

The process of the dewetting transition of amphiphile-stabilized liquid–liquid interfaces can be discussed utilizing the **standard model for depletion forces**. This model is based on hard colloidal particles surrounded by so-called

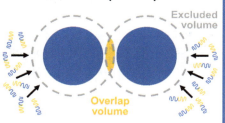

depletants, e.g., solvated polymers or micelles. If the excluded volume – a region surrounding each particle that cannot be occupied by depletants – of neighboring particles overlap, **depletants are expelled from the inter-particle space**. Thus, the **depletant concentration is higher in the surrounding volume than between the particles, leading to an osmotic pressure and attractive forces driving the particles together**. Similarly, O/W and W/O interfaces – instead of hard particles – are attracted due to excess amphiphile molecules acting as depletants in-between these layers. Depletion forces drive the amphiphile single layers towards each other like a zipper, leading to the dewetting of double emulsions.

As we continue along the double emulsion-to-vesicle transition process, the dewetted equilibrium structure with the organic phase attached to the as-formed vesicle will undergo a final transition. Either, the organic solvents remain attached to the as-formed vesicle surface and dry out, or more ideally, the organic solvent droplet – containing excess amphiphiles depending on their initial concentration used for the emulsion-template formation – completely detaches from the vesicle surface after the dewetting transition to yield a defect-free vesicle structure (Figure 2.18). The latter process can be observed just occasionally, probably induced by local shear forces during the collection of the emulsion templates and vesicles, respectively, or induced by structural imperfections at the contact area of the vesicle bilayer and the organic-solvent droplet.[7]

Up to this point, the focus has been on spherical vesicles prepared from rather simple double-emulsion templates with only one aqueous core. The power of microfluidics, however, lies in its ability to form emulsion droplets with much more complex architectures (cf. Figure 2.2C). For instance, by co-encapsulating two aqueous droplets in the organic-solvent phase of a W/O/W double emulsion, upon a dewetting transition, **double-core vesicles with two aqueous compartments** are obtained, whose composition can be individually tailored, for example, for dual-drug encapsulation, delivery, and release in combined therapy [80, 84]. Beyond the number of vesicle compartments, we are able to **control the vesicle size and internal content as well as the membrane composition, elasticity, and thickness by the molecular structure of the amphiphile building blocks** (e.g., by their hydrophobic block length). Owing to this myriad of parameters that we can master, the application of vesicles made by droplet microfluidics has extended in recent years from simple materials design toward emerging fields such as synthetic biology, which is devoted toward mimicking and understanding cellular processes.

---

[7] Quantitative detachment of the organic-solvent phase can be observed if the inner aqueous phase of amphiphile-stabilized double emulsions is solidified before the dewetting transition [88]. For example, by encapsulating monomers or prepolymers into the inner aqueous phase, solidification can yield polymer microgels. As the double emulsion droplets with a solidified core are transferred into an amphiphile-free collection phase, the lack of amphiphiles in this environment causes a marked change of the interfacial tension, which destabilizes the organic-solvent droplet surrounding the solidified core and causes its complete separation from the inner aqueous phase. Potential adherence of excess amphiphiles on the surface of the solidified particle leading to structural anisotropy is then not an issue. As a side note, if we employ polymerizable amphiphiles for stabilization of a W/O/W double emulsion, solidification of the droplet interface will yield amphiphilic capsules. Thus, by suitable choice of the amphiphile, W/O/W double emulsions can serve as a versatile platform for fabricating vesicles, capsules, and even particles.

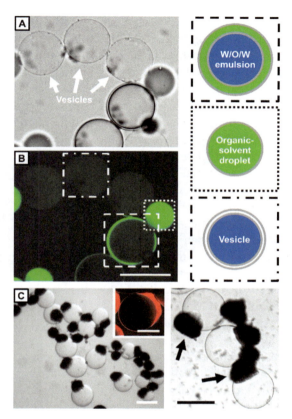

Figure 2.18: Vesicles made from double emulsion-based templates after undergoing a dewetting transition. Bright-field microscopy (A) and confocal fluorescence microscopy (B) images of three major species that can be found in a sample collected from a microfluidic device: vesicles, double emulsions, and organic-solvent droplets.[8] Excess amphiphile has been incorporated into the vesicle membrane during the emulsion-to-vesicle transition, as indicated by the coarse surface of the vesicle bilayer (marked by arrows). The scale bar that applies to both (A) and (B) denotes 50 μm. Adapted from [89]; copyright 2014 Royal Society of Chemistry. (C) Vesicles with excess amphiphile attached to their membrane. The hydrophobic bilayer and the excess-amphiphile patch of each vesicle are visualized by addition of Nile Red – a hydrophobic fluorescent dye – to the organic-solvent phase during the emulsion preparation (inset figure). The scale bars in both main panels denote 50 μm and 20 μm for the inset figure.

---

8 Visualizing vesicles by both bright-field and wide-field epifluorescence microscopy can be challenging: the vesicle bilayer, which is only nanometers thick even if excess amphiphile molecules have been trapped within the membrane during fabrication, is nearly index-matched with the surrounding aqueous media. A suitable method for visualizing vesicles and their membrane is thus confocal laser-scanning or confocal spinning-disc microscopy.

In a spectacular branch of application, droplet-templated vesicles can serve to mimic and model natural cellular functions. One of these is vesicle budding, which cells utilize in intra- and intercellular communication and for protecting sensitive goods against (enzymatic) degradation. To mimic vesicle budding at the surface of emulsion-templated vesicles, the amphiphile concentration needs to be adequate to cover and stabilize the droplet templates' interfaces, and to provide additional building blocks for the formation of the vesicle buds. If the initial concentration of amphiphiles is sufficiently high, excess amphiphile molecules that are not part of the bilayer will be trapped inside the vesicle membrane, as the inner-to-middle and middle-to-outer droplet interface of the droplet template close up. Followed by hydration and swelling of the vesicle bilayer by the surrounding aqueous phase, excess amphiphile molecules self-assemble inside the swollen vesicle membrane by lateral diffusion into a secondary vesicle bilayer, which then evaginates from the original vesicle membrane, occasionally followed by complete detachment from the outer vesicle surface [89]. Although vesicles made from lipids or copolymers seem to be a rather simple reflection of a complete living cell, they actually provide a straightforward path to mimic such an intricate cellular function like bud formation just by simply adjusting the amphiphile concentration, as this example demonstrates.

In another example, vesicles may be loaded with all necessary components to resemble the synthesis of ATP – the fuel of cellular life – on this account generating artificial mitochondria that can be useful as artificial bioinspired power source, for instance [90]. Other examples include the encapsulation of a bacteria's gene expression machinery for protein synthesis in vesicles [91], and even the large number of individual proteins involved in force generation systems as required for cellular division can be orchestrated on micrometer lengthscales inside droplet-templated vesicles [92, 93]. All these examples (cf. Figure 2.19) direct toward the ultimate application of vesicles prepared by droplet microfluidics in synthetic biology, which is the design of artificial cells to recreate and understand natural cellular functions and structures in a bottom-up approach [94, 95].

2.4.4 Droplet sizes and limitations

SF microfluidics typically deals with surfactant-stabilized droplets in the size range of 5–500 µm; these droplets are versatile platforms and precursors for further syntheses, material templating, and analytics. However, due to the minimal feature size of microflow cells that can be achieved by conventional combined photo- and soft lithography utilizing photomasks (cf. Sections 3.1.4 and 3.1.5), the **smallest microfluidic droplet size that we obtain is usually as small as 5 µm**, too. On top of this practical

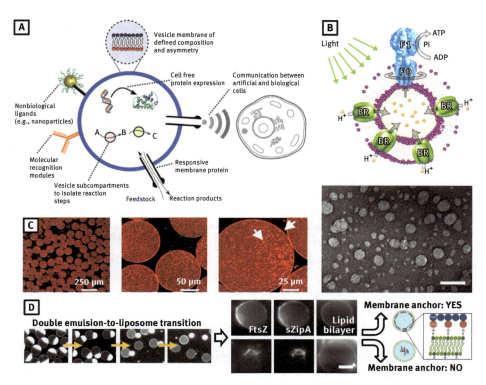

Figure 2.19: Vesicles for constructing and mimicking cellular functions as well as compartments. (A) Schematic of a vesicle-based approach to design artificial cells. Reproduced from [95]; copyright 2016 Portland Press Ltd. (B) Schematic of a polymersome, whose membrane incorporates both a photoactive, energy-transducing protein (bacteriorhodopsin, BR) and an ATP synthase. By coupling the action of these two proteins in one compartment, a proton gradient formed by the BR inside the polymersome fuels the synthesis of ATP from ADP and inorganic phosphate. The scale bar is 100 nm. Reproduced from [90]; copyright 2005 American Chemical Society. (C) Cell-free protein expression in polymersomes. Arrows indicate agglomeration of proteins that are formed by the droplet-encapsulated gene expression machinery. Reproduced from [91]; copyright 2012 WILEY-VCH. (D) Liposome formation from double emulsions undergoing a dewetting transition (top row). The middle and lower rows of images show the localization of FtsZ and sZipA proteins – naturally involved in bacterial cell division – encapsulated inside liposomes. In the presence of anchor lipids in the membrane, a circle-like assembly of the proteins is observed (middle row); in the absence of anchor points, bundle and filament formation is observed (bottom row). The scale bar for the lower group of images denotes 10 μm. Reproduced from [92]; copyright 2016 Nature Publishing Group.

limitation, thereare two additional fundamental limits for the minimal microfluidic droplet size. First, as we have seen in Section 1.5.3, operating a SF microfluidic device in the controlled dripping regime requires the fluid to flow at low Ca and We numbers, that is, at **low linear flow speed**. At a given volumetric flow rate, the linear flow speed scales with the inverse square of the microchannel diameter, meaning that the volumetric flow rate must be scaled down with a power of two if the channel diameter shall be made smaller such to keep Ca and We fixed. This, however, means that the droplet production frequency will dramatically decrease if the channel width shall be smaller. Second, even if we accept the latter limitation, the Hagen–Poiseuille law (eq. 1.17 in conjunction with Figure 1.4) tells us that the pressure that is needed to press a fluid through a microchannel scales with the channel width to the power of −4. That means that **the smaller the channel shall be, the extraordinarily larger the pressure must be to feed it with fluid, which will soon reach practical limits**. These inherent principles delimit the use of microfluidic channels pretty much to be larger than about 5 μm. Formation of **smaller droplets and particles thus requires rethinking of the experimental design** in droplet microfluidics. As one idea, microflow cells with flow-focusing junctions can be operated in a high jetting regime forming a thin thread of the dispersed phase that breaks into droplets much smaller than the microchannel diameter and minimal feature size of the flow cell, respectively. Alternatively, the liquid continuous phase can be exchanged by a flow of compressed gas (e.g., air or nitrogen) to form a spray of microdroplets. While in conventional droplet microfluidics for materials design, microdroplets ideally transition into a particle of the same size, nebulization of droplets and extremely fast evaporation allow for preparing nanoparticles as small as 20 nm.

Based on these properties of microfluidic spray formation, its most outstanding application is the formulation of nanoparticles by evaporative precipitation of sprayed solutions, containing drugs or inorganic salts, for instance. As the sprayed droplets are of pico- to femtoliter volume, and as their evaporation in air at supersonic ejection speed is fast, drying is complete before crystallization of the solutes sets in. This way, nanoparticles formed in the process of evaporative precipitation by microfluidic spray drying are purely amorphous, which adds new material properties to known naturally crystalline compounds, even table salt. In the area of pharmaceutics and drug formulations, this may improve the performance of new chemical entities from industrial screening programs as well as existing drugs, which often exhibit poor bioavailability due to hydrophobic functional groups, for example, fluorine. By processing macroscopic crystalline materials – for example, hydrophobic steroids – into nanoscopic amorphous materials by microfluidic spray drying, superior bioavailability and uptake rates are expected compared to corresponding crystalline drug species [96, 97].

Take-home messages Chapter 2.4 (Part 1)

Emulsions as reaction vessels and templates

- Microdroplets are utilized as compartments for synthesis and analytics.
- Microfluidically prepared emulsions allow for single-cell / single molecule detection in a defined volume (nL to fL), whose content can be analyzed by mass spectrometry as wenn as fluorescence, Raman, and IR spectroscopies.
- Droplets can be applied as templates with defined volume and internal organization (Janus / core–shell) for polymer microparticle fabrication.
- Emulsions may be solidified on chip, thereby potentially arresting the droplets in a non-equilibrium shape, or off chip, thereby commonly yielding spherical solid objects.

Take-home messages Chapter 2.4 (Part 2)

Emulsions as reaction vessels and templates

- Due to mild / bio-orthogonal solidification strategies and control over a large range of physicochemical / mechanical properties, microfluidically prepared polymer particles may also serve as tailored experimental platforms in cell biology (→ artificial ECM) and cell-free biotechnology (→ cell-like reaction environment).
- Emulsions may direct the assembly of amphiphiles towards giant single and multi-compartment vesicles with nearly quantitative encapsulation efficiency.

2.5 Numbering up

The microfluidic way of forming one drop at a time introduces extreme precision and uniformity. These characteristics have drawn particular interest in materials design, where a myriad of examples has shown the ability of microfluidics to fabricate nanoscale and microscale materials with identical physicochemical and mechanical properties among a population. Examples include monodisperse polymer microparticles and capsules loaded with drugs, food, cosmetics, dyes, or other active compounds, which release their content by external triggers such as heat, pH, light, or magnetic fields. However, while milliliter-to-femtoliter sample consumption may be beneficial for processing valuable materials that are merely available in small quantitates – drugs at their early stage of development, for instance – this is **orders of magnitude smaller than industrial production scale of fine chemicals**, and thus a **major obstacle toward industrial utilization of microfluidics**. Industry requires at least liter-per-hour product output to translate a microflow cell-based technology into a profitable application. For instance, when employing a single microfluidic W/O drop-maker with rectangular flow-focusing geometry (diameter: 100 μm) in which a Newtonian liquid is processed into droplets at flow rates of 1,000 μL h^{-1} for the dispersed phase and 3,000 μL h^{-1} for the continuous phase, we require 250 microflow cells to fabricate 1 L of 100-μm emulsion droplets per hour. Aside from the challenge to manufacture so many flow cells without defects in a reproducible fashion, given the multitude of process steps of conventional flow cell design (cf. Chapter 3 for details), each flow cell also requires a set of fluid pumps, rendering this approach toward scaling-up microfluidic production rates time-consuming and costly. As a result, **commercialization based on microflow cells is hardly feasible**. In addition, realization of large-scale microfluidic processes highly depends on addressing at least one so-called killer application that cannot be realized by any other available technology and that addresses a driving demand in an emerging area, such as those of materials design, analytics, or energy conversion [98]. While the formation of monodisperse single emulsions and homogeneous microparticles, respectively, in the size range of approx. 100 μm is the least challenging microfluidic experiment to be scaled-up, industrial fabrication with common emulsification technology is already feasible to do the same at ton-scales per year and with particle size polydispersities of sub-10%. A replacement of this industrial-scale materials design by microfluidic processing for the sake of just a slightly improved particle size distribution is thus unconvincing. A powerful strength of microfluidics, however, lies in its ability to fabricate **complex materials** – particularly **capsules** with **core–shell structures** – with **quantitative encapsulation efficiency** and tailored shell properties for triggered decomposition. Such entities cannot be fabricated by conventional membrane emulsification, not even in a multistep fashion. As introduced in Section 2.4.2, capsule materials are highly desirable for targeted drug delivery and protection of volatile as well as precious compounds. Here, microfluidic processing – despite comparable low product output – can make an impact by adding value to existing materials, as several SME examples indeed indicate (among others: Calyxia, Capsum) [99]. Such capsules are also desired as high-performance additives in construction chemistry, for example, when triggered fluidization and solidification of concrete at a tailored pH is of interest. Capsules can also serve for improved storage and extended release of active ingredients in laundry detergents.

To get into an industrial focus for this kind of application, **microfluidics must be enhanced in throughput**. Naturally, the key to achieving this enhancement lies in **parallelization** of microfluidic processes, thereby **numbering-up** the throughput.[9] For this purpose, a device fabrication technique that continuously gains in importance in microfluidic parallelization is rapid prototyping based on fused deposition modeling (FDM) and stereolithography (SL). Due to the freedom in device design toward highly complex multilayered structures, low material cost, and short manufacturing time from a CAD file to a device, additive manufacturing allows for fabricating monolithic devices, which exemplarily produce 500-μm droplets from 28 ladder-type drop makers at liters-per-hour product output, in a single manufacturing step [100]. Yet, examples of parallelization utilizing additive manufacturing techniques are still limited. In view of the focus of this textbook, we look at numbering-up microfluidics based on the two still most common device types in R&D microfluidics: glass capillaries (cf. Section 2.3.1) and stamped flow cells (cf. Section 2.3.2).

Parallelization and high-throughput is nearly impossible to be achieved with glass-capillary microfluidic devices, as a set of glass capillaries has to be aligned individually and manually for each microfluidic drop maker. Although efforts have been made in parallelization, for example, by coaxial arrays of densely packed capillaries [101] that are fed with liquids by one set of inlet ports for the dispersed and the continuous phases (in case of W/O or O/W single-emulsion formation), this approach cannot be scaled-up indefinitely in its current state. For example, four devices operated in parallel via this method already lead to a variance in droplet size distribution close to 10%.

In contrast to the limited prospect for glass-capillary devices, **lithography-based fabrication of elastomeric (e.g., PDMS-based) channel networks allows multiple**

---

[9] You may ask, why numbering up is not in the focus of continuous-flow microfluidics, but merely discussed for SF microfluidics. Indeed, low product output of a single microfluidic device is a key challenge in both continuous-flow and SF microfluidics. However, due to high-precision Computerized Numerical Control (CNC) machining as well as laser ablation processing, solvent- and high pressure-resistant as well as easy-to-clean flow reactors made of durable steel are already available with micrometer-scale precision and have been integrated into larger, parallelized assemblies for continuous-flow spray or co-flow formation. A key example is the formulation of carotenoids into nanoparticles for food industry via spray drying in continuous-flow-mode with tailored microscopic nozzles. While this and other examples demonstrate the feasibility of up-scaling continuous-flow microfluidics, similar examples for complex emulsions made by SF microfluidics are rare, and even upscaling a relatively simple W/O/W or O/W/O emulsion is a major challenge. Thus, this chapter discusses attempts to overcome the limitations of SF microfluidics toward industrial fluid-flow processing.

tens and hundreds of nearly identical channels to be parallelized, thereby allowing the production of single- [102] or multiple-emulsion [103] droplets to be scaled-up while maintaining the inherent advantage of exquisite droplet monodispersity and uniformity. To get to this point, several design rules for parallelized microfluidic devices have to be considered. While both T- and flow-focusing cross-junctions (Figure 2.20A) have been utilized for parallelized droplet formation, an even distribution of the microflow liquids to these drop makers is crucial to ensure stability of device operation, which could affect the droplet size and its dispersity [104]. Here, fluid flow simulations indicate that a ladder-like arrangement of individual drop makers (Figure 2.20B) provides superior flow stability over a tree-shaped architecture (Figure 2.20C) in the presence of defects, which are likely to occur facing the sheer number of drop makers that are integrated in parallelized devices. As fluid flow from one manifold is distributed via multiple branches, channel-size variations, dust particles, or partial microchannel clogging in each branch can lead to an increasing error in fluid flow distribution at each additional branching operation.

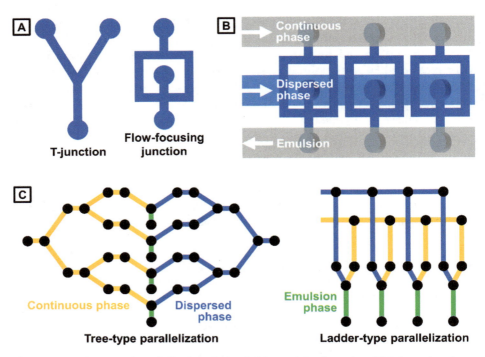

Figure 2.20: Schematic of parallelization of microfluidic emulsion formation. (A) Schematics of common drop makers with different inflow channel geometries. (B) Multilayered design of a stamped microflow cell with large channels in the top layer feeding three flow-focusing cross-junctions that operate in parallel in the bottom layer. (C) Prediction of the best microchannel designs for the numbering up of T-shaped drop makers.

Beyond the need for an even distribution of the fluid flow to all parallelized drop makers, a **simultaneous startup of all drop makers is crucial to ensure emulsion uniformity**. However, as liquids are flushed into a ladder-type microfluidic device (from left to right in Figure 2.20C), the flow will take the path of least resistance and bypass the drop maker array before all channels are filled in the back of the device to initiate the emulsion formation. By including drainages behind the drop maker array in the upper-layer inflow channels, the parallelized flow cell can be efficiently loaded with fluids before emulsion formation is initiated simultaneously in all drop makers by mechanical closure of the drainages. Lastly, it is important to consider the flow behavior once the device is in operation. As the outflow channel fills with droplets from all connected drop makers, these droplets cause a flow resistance that is different from that of the pure continuous phase. Thus, if the outflow channel is too narrow, backpressure or even clogging of the microchannel can lead to a feedback that slows down and eventually inhibits further droplet formation (Figure 2.21).

To ensure stable emulsion formation independent of the filling state of the flow cell, reservoir channels with a large opening should be included that are five to ten times larger in diameter than the cross-section of the droplet-forming nozzles.

Based on these design criteria, elastomeric PDMS devices containing 1,000 parallel flow-focusing devices arranged in a ladder geometry have been developed for forming single emulsions (45 µm in diameter) with liter-per-hour product output and low droplet size dispersity (approx. ±6%) [105]. Alternatively, mass production of monodisperse single emulsions has been implemented in **disc-like flow cells** with stackable layers of distribution channels, drop generators, and collection reservoirs [106]. Exemplarily, these stacked devices, integrating 128 flow-focusing junctions, form droplets (200 µm in diameter) at liter-per-hour rates and droplet size dispersities of ±6%, similar to the ladder-type flow-focusing junctions just discussed above. Potentially, this approach might be extendable toward forming multiple emulsions by employing stacks made from different layer materials. For instance, W/O droplets formed in a hydrophobic layer could feed into a second set of flow-focusing junctions inside a hydrophilic layer to produce W/O/W double emulsions.

As introduced previously, monodisperse emulsion droplets made by microfluidics are useful as templates for polymer microgels, cell mimics, or as reaction vessels in academic research. However, for industry, there is hardly any economic benefit to replace existing large-scale emulsification processes by numbering up droplet microfluidics. This situation is radically different for the use of microfluidics to forming capsules and higher-order emulsions, respectively. On this account, double-flow-focusing junctions for forming W/O/W and O/W/O emulsions have been aligned in a ladder-like architecture. Operated in a one-step emulsification regime to circumvent the challenge of synchronizing the droplet formation in two consecutive junctions (cf. Section 2.2.4), that kind of PDMS-based flow cell is able to produce approx. 1 kg per day of 150 µm-sized double emulsions with a size dispersity

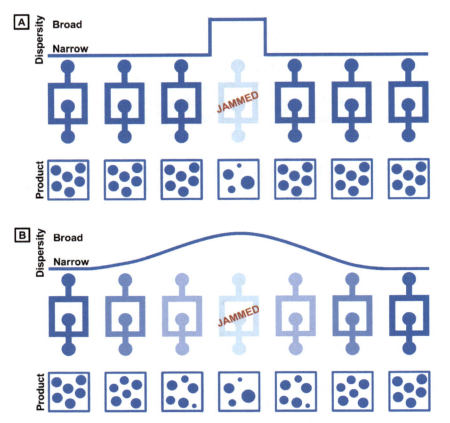

Figure 2.21: Jamming of a drop maker and its consequences for droplet formation in neighboring devices using the example of parallelized flow-focusing cross-junctions. (A) Ideally, as a single device jams due to clogging, for instance, droplet formation in all neighboring devices should be unaffected. (B) In fact, however, a pressure feedback caused by a jammed device can propagate to the neighboring drop makers and disturb the droplet breakoff and formation frequency on those as well, the latter one being particularly crucial in synchronized droplet formation of consecutive flow-focusing junctions.

of less than ±6% from 15 parallelized individual flow-focusing devices [107]. A key feature of this device is that it is fabricated in a **nonplanar fashion**. In such devices, microchannels do not solely extend in the x–y plane, but they also exhibit variations in the z-direction, as has been discussed at the end of Section 2.3.3 before. For instance, a flow-focusing junction may consist of three inflow channels in one plane, and an outflow channel that additionally extends above and below this plane. In such nonplanar droplet-forming microflow cells, the dispersed phase is fully surrounded by a protective sheath flow of the continuous phase. This way, the inner phase cannot make contact with the microchannel wall, such that accidental wetting of the inner droplet phases at the microchannel walls is prevented, and a spatial wettability pattern is thus not required. To fabricate nonplanar microflow cells by established photolithography, however, so-called mask aligners are required. These devices allow for precisely aligning several photomasks to yield a nonplanar lithography stamp for further replication of a nonplanar microchannel network in an elastomer. However, mask aligners are costly, and many users do not have access to them. In these cases, numbering up microfluidics must be implemented utilizing planar flow cells. Yet, simultaneously controlling the spatial wettability for multiple-emulsion formation – by LbL assembly or grafting-to or grafting-from polymerization of hydrophilic polymers, for instance – is challenging in such drop makers, especially if they are parallelized in an up-numbered array. A simple, yet efficient method to control the device wettability in a spatially resolved fashion is by **localized oxygen-plasma treatment** [26]. The plasma discharge cloud renders the surface of microchannels in PDMS hydrophilic (cf. Section 3.4.1), and it can be confined to defined areas of the flow cell by simply selectively blocking its fluid inflow and outflow ports of those parts that shall not be plasma-treated, thereby hindering the oxygen plasma to diffuse into the microchannel network at the blocked ports. Although hydrophilicity is only temporary in this case, as the plasma-induced surface modification yields just labile silanol groups, such a microflow cell can be employed for numbering up double-emulsion formation for at least a period of hours to days.

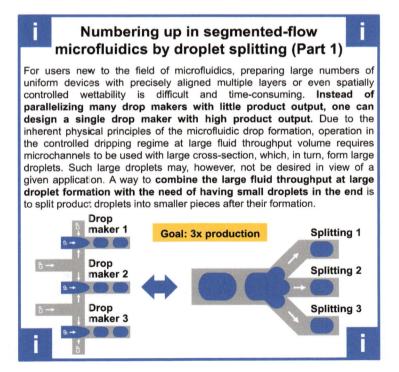

Numbering up in segmented-flow microfluidics by droplet splitting (Part 1)

For users new to the field of microfluidics, preparing large numbers of uniform devices with precisely aligned multiple layers or even spatially controlled wettability is difficult and time-consuming. **Instead of parallelizing many drop makers with little product output, one can design a single drop maker with high product output.** Due to the inherent physical principles of the microfluidic drop formation, operation in the controlled dripping regime at large fluid throughput volume requires microchannels to be used with large cross-section, which, in turn, form large droplets. Such large droplets may, however, not be desired in view of a given application. A way to **combine the large fluid throughput at large droplet formation with the need of having small droplets in the end** is to split product droplets into smaller pieces after their formation.

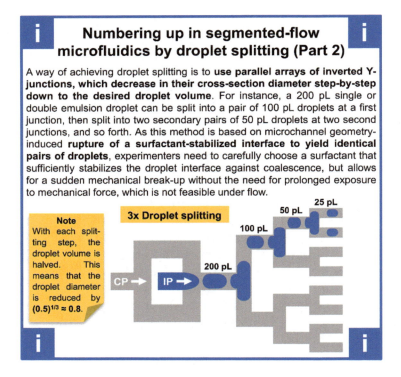

To further discuss microfluidic scale-up, let us focus on the operation of these devices. Conventionally, parallelized microfluidic devices are operated by active fluid-flow control employing syringe pumps. By setting a defined ratio of flow rates – for example, 3:1 for the continuous:dispersed phases in the case of Newtonian fluids – we ensure controlled droplet breakup in a dripping regime. Yet, fluctuations of the pumps or clogging of neighboring junctions within an array of microchannels compromises the narrow size distribution of the emulsions obtained, thereby losing one of the key selling points of droplet microfluidics. To overcome this challenge, **purely geometry-controlled droplet breakup** provides a means to operate parallelized flow cells in a robust fashion independent of flow fluctuations, even if the fluids are injected via syringes just by hand instead of high-precision pumps. This method relies on a static instability that pinches-off droplets from a fluid reservoir that gradually fills with liquid before the next droplet pinches-off. Massive parallelization of this geometry-based droplet formation (550 nozzles) therefore allows for preparing emulsion droplets with sizes between 20 µm and 200 µm at rates well above liters per hour and with significantly lower size polydispersity than that achieved with scaled-up flow-focusing cross- and Y-junction devices [108]. This approach has recently been extended toward double emulsion preparation yielding droplets with

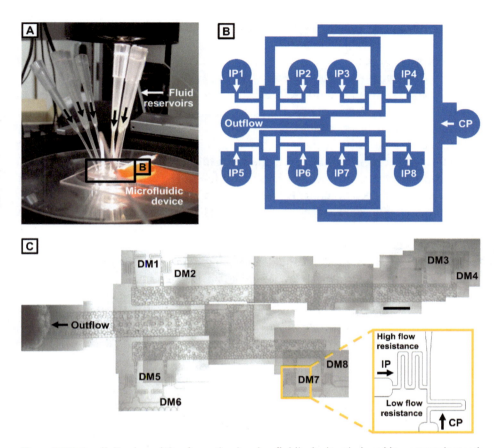

Figure 2.22: Parallelized emulsion formation in microfluidic devices induced by vacuum instead of pumps. (A) Photograph of an experimental setup consisting of a stamped flow cell connected to pipette tips with flexible head as used for gel loading in gel electrophoresis. From these fluid reservoirs, liquid content is dragged into the microfluidic device by vacuum, for example, applied by a syringe connected to the outflow channel. (B) Schematic of a microchannel design for fabricating microemulsions from eight individual microfluidic T-junctions operated in parallel. A continuous phase (CP) is distributed from one major inflow channel to the drop maker, which are fed from individual fluid reservoirs with different inner phases (IP). Such a microfluidic device allows for simultaneously forming emulsions with eight different compositions, as required for screening experimental parameters in (bio-)chemical microsynthesis and analysis, for instance. (C) Composite bright-field microscopy image of the device in (B) (DM = drop maker). In the absence of an active fluid pumping system, the hydrodynamic resistance of the microchannels for the dispersed phases and the continuous phase (figure inset) determines the optimal flow-rate ratio required for stable droplet formation. The scale bar denotes 500 µm.

extremely thin shells (down to 240 nm) in a high-throughput fashion (1,000 droplets per second) and at a high volume fraction (up to 95% v/v) [109].

From such robust, geometry-controlled droplet breakup using simple instead of high-precision syringe pumps, the next step in simplifying parallelized (higher-order) emulsion formation is to **completely remove any bulky pumps and computers to control the flow pattern formation** in parallelized flow cells [110]. Instead of actively pumping fluids – conventionally from the inflow channel – into a flow cell, we may move fluids through a microchannel network by applying a **vacuum at the outflow channel**, which drags fluids (e.g., a defined pair of continuous and dispersed phases) from simple reservoirs into the device, as shown in Figure 2.22A. In this example, a key parameter that we need to control is the flow-rate ratio of the dispersed and continuous phase for droplet breakup at each of the individual microchannel junctions. In the absence of an active pumping system for tailoring this parameter, it is helpful to **treat microchannels like electronic circuits** for microchannel design considerations. For controlling the flow-rate ratio, we particularly focus on electronic resistors and their microfluidic counterparts, which are meander-like channels (cf. Section 3.1.2 for more details). Here, the length of a meandering microchannel controls its **hydrodynamic resistance**. By utilizing a set of two meanders with dissimilar length and meander turns, as shown in Figure 2.22C, we can set a suitable flow-rate ratio of the dispersed phase and the continuous phase at the emulsion-forming junction.

Take-home messages Chapter 2.5

From microliters towards cubic meters

- Single-emulsion formation with narrow size distribution has been implemented on industrial scale without microfluidic systems. Yet, upscaling of higher-order emulsion formation is a major challenge.
- Large-scale formation of higher-order emulsions by droplet microfluidics is hampered by the low product output of individual flow cells.
- Upscaling droplet microfluidics towards industrial production rates can be achieved by parallelization, such that production rates scale linearly with the number of individual microfluidic systems combined to arrays.
- Most examples utilize stamped flow cells based on PDMS replicas.
- Interconnected, parallelized microfluidic systems may suffer from pressure feedbacks and flow fluctuations, requiring additional design features (e.g., dampening fluid reservoirs).
- Simplification of intrinsically complex parallelized microfluidic systems for higher-order emulsion formation is achieved by circumventing surface patterning (→ non-planar flow cells) and relying on passive, geometry-driven fluid distribution (→ no active pumps).

References

[1] S.-Y. Teh, R. Lin, L.-H. Hung, A. P. Lee, "Droplet microfluidics", *Lab Chip* 2008, *8*, 198. DOI: **10.1039/b715524g**.

[2] R. Seemann, M. Brinkmann, T. Pfohl, S. Herminghaus, "Droplet based microfluidics", *Rep. Prog. Phys.* 2012, *75*, 016601. DOI: **10.1088/0034-4885/75/1/016601**.

[3] G. Niu, A. Ruditskiy, M. Varac, Y. Xia, "Toward continuous and scalable production of colloidal nanocrystals by switching from batch to droplet reactors", *Chem. Soc. Rev.* 2015, *44*, 5806. DOI: **10.1039/c5cs00049a**.

[4] L. Rayleigh, "On the instability of a cylinder of viscous liquid under capillary force", *Philos. Mag.* 1892, *34*, 145. DOI: **10.1080/14786449208620301**.

[5] T. Y. Lee, T. M. Choi, T. S. Shim, R. A. M. Frijns, S.-H. Kim, "Microfluidic production of multiple emulsions and functional microcapsules", *Lab Chip* 2016, *16*, 3415. DOI: **10.1039/c6lc00809g**.

[6] H. Song, M. R. Bringer, J. D. Tice, C. J. Gerdts, R. F. Ismagilov, "Experimental test of scaling of mixing by chaotic advection in droplets moving through microfluidic channels", *Appl. Phys. Lett.* 2003, *83*, 4664. DOI: **10.1063/1.1630378**.

[7] H. Song, J. D. Tice, R. F. Ismagilov, "A microfluidic system for controlling reaction networks in time", *Angew. Chem. Int. Ed.* 2003, *42*, 767. DOI: **10.1002/anie.200390203**.

[8] M. R. Bringer, C. J. Gerdts, H. Song, J. D. Tice, R. F. Ismagilov, "Microfluidic systems for chemical kinetics that rely on chaotic mixing in droplets", *Philos. Transact. A Math. Phys. Eng. Sci.* 2004, *362*, 1087. DOI: **10.1098/rsta.2003.1364**.

[9] L. Jiang, Y. Zeng, H. Zhou, J. Y. Qu, S. Yao, "Visualizing millisecond chaotic mixing dynamics in microdroplets: A direct comparison of experiment and simulation", *Biomicrofluidics* 2012, *6*, 012810. DOI: **10.1063/1.3673254**.

[10] A. R. Abate, T. Hung, P. Mary, J. J. Agresti, D. A. Weitz, "High-throughput injection with microfluidics using picoinjectors", *Proc. Natl. Acad. Sci. USA* 2010, *107*, 19163. DOI: **10.1073/pnas.1006888107**.

[11] D. R. Link, S. L. Anna, D. A. Weitz, H. A. Stone, "Geometrically mediated breakup of drops in microfluidic devices", *Phys. Rev. Lett.* 2004, *92*, 054503. DOI: **10.1103/PhysRevLett.92.054503**.

[12] A. R. Abate, D. A. Weitz, "Faster multiple emulsification with drop splitting", *Lab Chip* 2011, *11*, 1911. DOI: **10.1039/C0LC00706D**.

[13] K. Ahn, J. J. Agresti, H. Chong, M. Marquez, D. A. Weitz, "Electrocoalescence of drops synchronized by size-dependent flow in microfluidic channels", *Appl. Phys. Lett.* 2006, *88*, 264105. DOI: **10.1063/1.2218058**.

[14] X. Niu, S. Gulati, J. B. Edel, A. J. deMello, "Pillar-induced droplet merging in microfluidic circuits", *Lab Chip* 2008, *8*, 1837. DOI: **10.1039/B813325E**.

[15] C. Holtze, A. C. Rowat, J. J. Agresti, J. B. Hutchison, F. E. Angile. C. H. J. Schmitz, S. Koester, H. Duan, K. J. Humphry, R. A. Scanga, J. S. Johnson, D. Pisignanoc, D. A. Weitz, "Biocompatible surfactants for water-in-fluorocarbon emulsions", *Lab Chip* 2008, *8*, 1632. DOI: **10.1039/b806706f**.

[16] J. J. Agresti, E. Antipov, A. R. Abate, K. Ahn, A. C. Rowat, J. C. Baret, M. Marquez, A. M. Klibanov, A. D. Griffiths, D. A. Weitz, "Ultrahigh-throughput screening in drop-based microfluidics for directed evolution", *Proc. Natl. Acad. Sci. USA* 2010, *107(9)*, 4004. DOI: **10.1073/pnas.0910781107**.

[17] A. R. Abate, D. A. Weitz, "High-order multiple emulsions formed in poly(dimethylsiloxane) microfluidics", *Small* 2009, *5*, 2030. DOI: **10.1002/smll.200900569**.

[18] I. Ziemecka, V. van Steijn, G. J. M. Koper, M. T. Kreutzer, J. H. van Esch, "All-aqueous core-shell droplets produced in a microfluidic device", *Soft Matter* 2011, *7*, 9878. DOI: **10.1039/c1sm06517c**.

[19] Y. Song, H. C. Shum, "Monodisperse W/W/W double emulsion induced by phase separation", *Langmuir* 2012, *28*, 12054. **DOI: 10.1021/la3026599**.
[20] A. S. Utada, E. Lorenceau, D. R. Link, P. D. Kaplan, H. A. Stone, D. A. Weitz, "Monodisperse double emulsions generated from a microcapillary device", *Science* 2005, *308*, 537. **DOI: 10.1126/science.1109164**.
[21] A. R. Abate, J. Thiele, D. A. Weitz, "One-step formation of multiple emulsions in microfluidics", *Lab Chip* 2011, *11*, 253. **DOI: 10.1039/c0lc00236d**.
[22] A. S. Utada, A. Fernandez-Nieves, H. A. Stone, D. A. Weitz, "Dripping to jetting transitions in coflowing liquid streams", *Phys. Rev. Lett.* 2007, *99*, 094502. **DOI: 10.1103/PhysRevLett.99.094502**.
[23] Y. Xia, G. M. Whitesides, "Soft Lithography", *Angew. Chem. Int. Ed.* 1998, *3*, 550. **DOI: 10.1002/(SICI)1521-3773(19980316)37:5<550::AID-ANIE550>3.0.CO;2-G**.
[24] J. C. McDonald, D. C. Duffy, J. R. Anderson, D. T. Chiu, H. K. Wu, O. J. A. Schueller, G. M. Whitesides, "Fabrication of microfluidic systems in poly(dimethylsiloxane)", *Electrophoresis* 2000, *21*, 27. **DOI: 10.1002/(SICI)1522-2683(20000101)21:1<27::AID-ELPS27>3.0.CO;2-C**.
[25] W.-A. C. Bauer, M. Fischlechner, C. Abell, W. T. S. Huck, "Hydrophilic PDMS microchannels for high-throughput formation of oil-in-water microdroplets and water-in-oil-in-water double emulsions", *Lab Chip* 2010, *10*, 1814. **DOI: 10.1039/c004046k**.
[26] S. C. Kim, D. J. Sukovich, A. R. Abate, "Patterning microfluidic device wettability with spatially-controlled plasma oxidation", *Lab Chip* 2015, *15*, 3163. **DOI: 10.1039/c5lc00626k**.
[27] A. R. Abate, J. Thiele, M. Weinhart, D. A. Weitz, "Patterning microfluidic device wettability using flow confinement", *Lab Chip* 2010, *10*, 1774. **DOI: 10.1039/c004124f**.
[28] A. Rotem, A. R. Abate, A. S. Utada, V. Van Steijn, D. A. Weitz, "Drop formation in non-planar microfluidic devices", *Lab Chip* 2012, *12*, 4263. **DOI: 10.1039/c2lc40546f**.
[29] T. M. Tran, S. Cater, A. R. Abate, "Coaxial flow focusing in poly(dimethylsiloxane) microfluidic devices", *Biomicrofluidics* 2014, *8*, 016502. **DOI: 10.1063/1.4863576**.
[30] R. H. Cole, T. M. Tran, A. R. Abate, "Double emulsion generation using a Polydimethylsiloxane (PDMS) Co-axial flow focus device", *J. Vis. Exp.* 2015, *106*, e53516. **DOI: 10.3791/53516**.
[31] L. M. Fidalgo, G. Whyte, B. T. Ruotolo, J. L. P. Benesch, F. Stengel, C. Abell, C. V. Robinson, W. T. S. Huck, "Coupling microdroplet microreactors with mass spectrometry: Reading the contents of single droplets online", *Angew. Chem. Int. Ed.* 2009, *48*, 3665. **DOI: 10.1002/anie.200806103**.
[32] S. K. Küster, S. R. Fagerer, P. E Verboket, K. Eyer, K. Jefimovs, R. Zenobi, P. S. Dittrich, "Interfacing droplet microfluidics with matrix-assisted laser desorption/ionization mass spectrometry: Label-free content analysis of single droplets", *Anal. Chem.* 2013, *85*, 1285. **DOI: 10.1021/ac3033189**.
[33] J. Heinemann, K. Deng, S. C. C. Shih, J. Gao, P. D. Adams, A. K. Singh, T. R. Northen, "On-chip integration of droplet microfluidics and nanostructure-initiator mass spectrometry for enzyme screening", *Lab Chip* 2017, *17*, 323. **DOI: 10.1039/c6lc01182a**.
[34] A. R. Abate, T. Hung, R. A. Sperling, P. Mary, A. Rotem, J. J. Agresti, M. A. Weiner, D. A. Weitz, "DNA sequence analysis with droplet-based microfluidics", *Lab Chip* 2013, *13*, 4864. **DOI: 10.1039/C3LC50905B**.
[35] A. Rotem, O. Ram, N. Shoresh, R. A. Sperling, M. Schnall-Levin, H. Zhang, A. Basu, B. E. Bernstein, D. A. Weitz, "High-Throughput Single-Cell Labeling (Hi-SCL) for RNA-Seq using drop-based microfluidics", *PLoS ONE* 2015, *10*, e0116328. **DOI: 10.1371/journal.pone.0116328**.

[36] P. Kumaresan, C. James Yang, S. A. Cronier, R. G. Blazej, R. A. Mathies, "High-throughput single copy DNA amplification and cell analysis in engineered nanoliter droplets", *Anal. Chem.* 2008, *80*, 3522. **DOI: 10.1021/ac800327d.**

[37] D. Pekin, Y. Skhiri, J.-C. Baret, D. Le Corre, L. Mazutis, C. Ben Salem, F. Millot, A. El Harrak, J. B. Hutchison, J. W. Larson, D. R. Link, P. Laurent-Puig, A. D. Griffiths, V. Taly, "Quantitative and sensitive detection of rare mutations using droplet-based microfluidics", *Lab Chip* 2011, *11*, 2156. **DOI: 10.1039/C1LC20128J.**

[38] A. C. Hatch, J. S. Fisher, A. R. Tovar, A. T. Hsieh, R. Lin, S. L. Pentoney, D. L. Yangb, A. P. Lee, "1-Million droplet array with wide-field fluorescence imaging for digital PCR", *Lab Chip* 2011, *11*, 3838. **DOI: 10.1039/C1LC20561G.**

[39] J. Shuga, Y. Zeng, R. Novak, Q. Lan, X. Tang, N. Rothman, R. Vermeulen, L. Li, A. Hubbard, L. Zhang, R. A. Mathies, M. T. Smith, "Single molecule quantitation and sequencing of rare translocations using microfluidic nested digital PCR", *Nucl. Acids Res.* 2013, *41*, e159. **DOI: 10.1093/nar/gkt613.**

[40] D.-K. Kanga, M. M. Alia, K. Zhanga, E. J. Ponea, W. Zhao, "Droplet microfluidics for single-molecule and single-cell analysis in cancer research, diagnosis and therapy", *TrAC: Trends in Analytical Chemistry* 2014, *58*, 145. **DOI: 10.1016/j.trac.2014.03.006.**

[41] M. M. K. Hansen, L. H. H. Meijer, E. Spruijt, R. J. M. Maas, M. Ventosa Rosquelles, J. Groen, H. A. Heus, W. T. S. Huck, "Macromolecular crowding creates heterogeneous environments of gene expression in picolitre droplets", *Nature Nanotechnology* 2016, *11*, 191. **DOI: 10.1038/nnano.2015.243.**

[42] S. Mellouli, L. Bousekkine, A. B. Theberge, W. T. S. Huck, "Investigation of 'On Water' conditions using a biphasic fluidic platform", *Angew. Chem. Int. Ed.* 2012, *51*, 7981. **DOI: 10.1002/anie.201200575.**

[43] A. Chanda, V. V. Fokin, "Organic synthesis 'On Water' ", *Chem. Rev.* 2009, *109*, 725. **DOI: 10.1021/cr800448q.**

[44] S. Narayan, J. Muldoon, M. G. Finn, V. V. Fokin, H. C. Kolb, K. B. Sharpless, " 'On Water': Unique reactivity of organic compounds in aqueous suspension", *Angew. Chem.* 2005, *117*, 3339. **DOI: 10.1002/ange.200462883.**

[45] W.-L. Chou, P.-Y. Lee, C.-L. Yang, W.-Y. Huang, Y.-S. Lin, "Recent advances in applications of droplet microfluidics", *Micromachines* 2015, *6*, 1249. **DOI: 10.3390/mi6091249.**

[46] L. Shang, Y. Cheng, Y. Zhao, "Emerging droplet microfluidics", *Chem. Rev.* 2017, *117*, 7964. **DOI: 10.1021/acs.chemrev.6b00848.**

[47] P. Zhu, L. Wang, "Passive and active droplet generation with microfluidics: A review", *Lab Chip* 2017, *17*, 34. **DOI: 10.1039/C6LC01018K.**

[48] H. Jönsson, S. H. Andersson, "Droplet microfluidics – A tool for single cell analysis", *Angew. Chem. Int. Ed.* 2012, *51*, 12176. **DOI: 10.1002/anie.201200460.**

[49] E. Brouzes, "Droplet microfluidics for single-cell analysis", *Methods Mol. Biol.* 2012, *853*, 105. **DOI: 10.1007/978-1-61779-567-1_10.**

[50] M. Zagnoni, J. M. Cooper, "Droplet microfluidics for high-throughput analysis of cells and particles", *Methods Cell Biol.* 2011, *102*, 25. **DOI: 10.1016/B978-0-12-374912-3.00002-X.**

[51] T. A. Duncombe, A. M. Tentori, A. E. Herr, "Microfluidics: Reframing biological enquiry", *Nat. Rev. Mol. Cell Biol.* 2015, *16*, 554. **DOI: 10.1038/nrm4041.**

[52] D.-K. Kang, M. M. Ali, K. Zhang, E. J. Pone, W. Zhao, "Droplet microfluidics for single-molecule and single-cell analysis in cancer research, diagnosis and therapy", *TrAC: Trends in Analytical Chemistry* 2014, *58*, 145. **DOI: 10.1016/j.trac.2014.03.006.**

[53] S. Seiffert, "Functional microgels tailored by droplet-based microfluidics", *Macromol. Rapid. Commun.* 2011, *32*, 1600. **DOI: 10.1002/marc.201100342.**

[54] S. Seiffert, "Microgel capsules tailored by droplet-based microfluidics", *ChemPhysChem* 2013, *14*, 295. DOI: 10.1002/cphc.201200749.

[55] S. Seiffert, "Small but smart: Sensitive microgel capsules", *Angew. Chem. Int. Ed.* 2013, *52*, 11462. DOI: 10.1002/anie.201303055.

[56] S. Seiffert, D. A. Weitz, "Controlled fabrication of polymer microgels by polymer-analogous gelation in droplet microfluidics", *Soft Matter* 2010, *6*, 3184. DOI: 10.1039/C0SM00071J.

[57] J.-W. Kim, A. S. Utada, A. Fernndez-Nieves, Z. Hu, D. A. Weitz, "Fabrication of monodisperse gel shells and functional microgels in microfluidic devices", *Angew. Chem. Int. Ed.* 2007, *46*, 1819. DOI: 10.1002/anie.200604206.

[58] Y. Ma, J. Thiele, L. Abdelmohsen, J. Xu, W. T. S. Huck, "Biocompatible macro-initiators controlling radical retention in microfluidic on-chip photo-polymerization of water-in-oil emulsions", *Chem. Commun.* 2014, *50*, 112. DOI: 10.1039/c3cc46733c.

[59] S. Xu, Z. Nie, M. Seo, P. Lewis, E. Kumacheva, H. A. Stone, P. Garstecki, D. B. Weibel, I. Gitlin, G. M. Whitesides, "Generation of monodisperse particles by using microfluidics: Control over size, shape, and composition", *Angew. Chem. Int. Ed.* 2005, *44* 724. DOI: 10.1002/anie.200462226.

[60] T. Nisisako, T. Torii, "Formation of biphasic janus droplets in a microfabricated channel for the synthesis of shape-controlled polymer microparticles", *Adv. Mater.* 2007, *19*, 1489. DOI: 10.1002/adma.200700272.

[61] S. Ma, J. Thiele, X. Liu, Y. Bai, C. Abell, W. T. S. Huck, "Fabrication of microgel particles with complex shape via selective polymerization of aqueous two-phase systems", *Small* 2012, *8*, 2356. DOI: 10.1002/smll.201102715.

[62] T. Nisisako, T. Torii, T. Takahashi, Y. Takizawa, "Synthesis of monodisperse bicolored janus particles with electrical anisotropy using a microfluidic co-flow system", *Adv. Mater.* 2006, *18*, 1152. DOI: 10.1002/adma.200502431.

[63] K. P. Yuet, D. K. Hwang, R. Haghgooie, P. S. Doyle, "Multifunctional superparamagnetic janus particles", *Langmuir* 2010, *26*, 4281. DOI: 10.1021/la903348s.

[64] T. Tanaka, M. Okayama, H. Minami, M. Okubo, "Dual stimuli-responsive 'mushroom-like' Janus polymer particles as particulate surfactants", *Langmuir* 2010, *26*, 11732. DOI: 10.1021/la101237c.

[65] A. X. Lu, Y. Liu, H. Oh, A. Gargava, E. Kendall, Z. Nie, D. L. DeVoe, S. R. Raghavan, "Catalytic propulsion and magnetic steering of soft, patchy microcapsules: Ability to pick-up and drop-off microscale cargo", *ACS Appl. Mater. Interfaces* 2016, *8*, 15676. DOI: 10.1021/acsami.6b01245.

[66] A. Engler, S. Sen, H. Sweeney, D. Discher, "Matrix elasticity directs stem cell lineage specification", *Cell* 2006, *126*, 677. DOI: 10.1016/j.cell.2006.06.044.

[67] Y. Ma, M. P. Neubauer, J. Thiele, A. Fery, W. T. S. Huck, "Artificial microniches for probing mesenchymal stem cell fate in 3D", *Biomater. Sci.* 2014, *2*, 1661. DOI: 10.1039/C4BM00104D.

[68] D. M. Headen, J. R. García, A. J. García, "Parallel droplet microfluidics for high throughput cell encapsulation and synthetic microgel generation", *Microsyst. Microeng.* 2018, *4*, 17076. DOI: 10.1038/micronano.2017.76.

[69] H. Zhang, E. Tumarkin, R. Peerani, Z. Nie, R. M. A. Sullan, G. C. Walker, E. Kumacheva, "Microfluidic production of biopolymer microcapsules with controlled morphology", *J. Am. Chem. Soc.* 2006, *128*, 12205. DOI: 10.1021/ja0635682.

[70] N. Vilanova, C. Rodríguez-Abreu, A. Fernández-Nieves, C. Solans, "Fabrication of novel silicone capsules with tunable mechanical properties by microfluidic techniques", *ACS App. Mater. Interfaces* 2013, *5*, 5247. DOI: 10.1021/am4010896.

[71] S.-H. Kim, J.-G. Park, T. M. Choi, V. N. Manharan, D. A. Weitz, "Osmotic-pressure-controlled concentration of colloidal particles in thin-shelled capsules", *Nat. Commun.* 2014, *3068*, 1. DOI: 10.1038/ncomms4068.

[72] H.-J. Oh, S.-H. Kim, J.-Y. Baek, G.-H. Seong, S.-H. Lee, "Hydrodynamic micro-encapsulation of aqueous fluids and cells via 'on the fly' photopolymerization", *J. Micromech. Microeng.* 2006, *16*, 285. **DOI: 10.1088/0960-1317/16/2/013**.

[73] L. Martin-Banderas, M. Flores-Mosquera, P. Riesco-Chueca, A. Rodriguez-Gil, A. Cebolla, S. Chávez, A. M. Gañán-Calvo, "Flow focusing: A versatile technology to produce size-controlled and specific-morphology microparticles", *Small* 2005, *1*, 688. **DOI: 10.1002/smll.200500087**.

[74] S. Seiffert, J. Thiele, A. R. Abate, D. A. Weitz, "Smart microgel capsules from macromolecular precursors", *J. Am. Chem. Soc.* 2010, *132*, 6606. **DOI: 10.1021/ja102156h**.

[75] J. Thiele, S. Seiffert, "Double emulsions with controlled morphology by microgel scaffolding", *Lab Chip* 2011, *11*, 3188. **DOI: 10.1039/c1lc20242a**.

[76] J.-T. Wang, J. Wang, J.-J. Han, "Fabrication of advanced particles and particle-based materials assisted by droplet-based microfluidics", *Small* 2011, *7*, 1728. **DOI: 10.1002/smll.201001913**.

[77] J. H. Kim, T. Y. Jeon, T. M. Choi, T. S. Shim, S.-H. Kim, S.-M. Yang, "Droplet microfluidics for producing functional microparticles", *Langmuir* 2014, *30*, 1473. **DOI: 10.1021/la403220p**.

[78] D. Lee, D. A. Weitz, "Double emulsion-templated nanoparticle colloidosomes with selective permeability", *Adv. Mater.* 2008, *20*, 3498. **DOI: 10.1002/adma.200800918**.

[79] H. C. Shum, J.-W. Kim, D. A. Weitz, "Microfluidic fabrication of monodisperse biocompatible and biodegradable polymersomes with controlled permeability", *J. Am. Chem. Soc.* 2008, *130*, 9543. **DOI: 10.1021/ja802157y**.

[80] H. C. Shum, Y.-j. Zhao, S.-H. Kim, D. A. Weitz, "Multicompartment polymersomes from double emulsions", *Angew. Chem. Int. Ed.* 2011, *50*, 1648. **DOI: 10.1002/anie.201006023**.

[81] J. Thiele, A. R. Abate, H. C. Shum, S. Bachtler, S. Förster, D. A. Weitz, "Fabrication of polymersomes using double-emulsion templates in glass-coated stamped microfluidic devices", *Small* 2010, *6*, 1723. **DOI: 10.1002/smll.201000798**.

[82] S.-H. Kim, H. C. Shum, J.-W. Kim, J.-C. Cho, D. A. Weitz, "Multiple polymersomes for programmed release of multiple components", *J. Am. Chem. Soc.* 2011, *133*, 15165. **DOI: 10.1021/ja205687k**.

[83] Y. Su, H. Zhao, J. Wu, J. Xu, "One-step fabrication of silica colloidosomes with *in situ* drug encapsulation", *RSC Adv.* 2016, *6*, 112292. **DOI: 10.1039/c6ra19048k**.

[84] N.-N. Deng, M. Yelleswarapu, W. T. S. Huck, "Monodisperse uni- and multicompartment liposomes", *J. Am. Chem. Soc.* 2016, *138*, 7584. **DOI: 10.1021/jacs.6b02107**.

[85] H. C. Shum, J.-W. Kim, D. A. Weitz, "Microfluidic fabrication of monodisperse biocompatible and biodegradable polymersomes with controlled permeability", *J. Am. Chem. Soc.* 2008, *130*, 9543. **DOI: 10.1021/ja802157y**.

[86] H. C. Shum, E. Santanach-Carreras, J.-W. Kim, A. Ehrlicher, J. Bibette, D. A. Weitz, "Dewetting-induced membrane formation by adhesion of amphiphile-laden interfaces", *J. Am. Chem. Soc.* 2011, *133*, 4420. **DOI: 10.1021/ja108673h**.

[87] R. C. Hayward, A. S. Utada, N. Dan, D. A. Weitz, "Dewetting instability during the formation of polymersomes from block copolymer-stabilized double emulsions", *Langmuir* 2006, *22*, 4457. **DOI: 10.1021/la060094b**.

[88] C.-H. Choi, H. Wang, H. Lee, J. H. Kim, L. Zhang, A. Mao, D. J. Mooney, D. A. Weitz, "One-step generation of cell-laden microgels using double emulsion drops with a sacrificial ultra-thin oil shell", *Lab Chip* 2016, *16*, 1549. **DOI: 10.1039/c6lc00261g**.

[89] J. Thiele, V. Chokkalingam, S. Ma, D. A. Wilson, W. T. S. Huck, "Vesicle budding from polymersomes templated by microfluidically prepared double emulsions", *Mater. Horiz.* 2014, *1*, 96. **DOI: 10.1039/c3mh00043e**.

[90] H.-J. Choi, C. D. Montemagno, "Artificial organelle: 'ATP synthesis from cellular mimetic polymersomes'", *Nano Lett.* 2005, *5*, 2538. **DOI: 10.1021/nl051896e**.

[91] C. Martino, S.-H. Kim, L. Horsfall, A. Abbaspourrad, S. J. Rosser, Jonathan Cooper, D. A. Weitz, "Protein expression, aggregation, and triggered release from polymersomes as artificial cell-like structures", *Angew. Chem. Int. Ed.* 2012, *51*, 6416. **DOI: 10.1002/anie.201201443.**

[92] S. Deshpande, Y. Caspi, A. E. C. Meijering, C. Dekker, "Octanol-assisted liposome assembly on chip", *Nat. Commun.* 2016, *7*, 10447. **DOI: 10.1038/ncomms10447.**

[93] E. J. Cabré, A. Sánchez-Gorostiaga, P. Carrara, N. Ropero, M. Casanova, P. Palacios, P. Stano, M. Jiménez, G. Rivas, M. Vicente, "Bacterial division proteins FtsZ and ZipA induce vesicle shrinkage and cell membrane invagination", *J. Biol. Chem.* 2013, *288*, 26625. **DOI: 10.1074/jbc.M113.491688.**

[94] S. Matosevic, "Synthesizing artificial cells from giant unilamellar vesicles: State-of-the art in the development of microfluidic technology", *Bioessays* 2012, *34*, 992. **DOI: 10.1002/bies.201200105.**

[95] Y. Elani, "Construction of membrane-bound artificial cells using microfluidics: A new frontier in bottom-up synthetic biology", *Biochem. Soc. Trans.* 2016, *44*, 723. **DOI: 10.1042/BST20160052.**

[96] J. Thiele, M. Windbergs, A. R. Abate, M. Trebbin, H. C. Shum, S. Förster, D. A. Weitz, "Early development drug formulation on a chip: Fabrication of nanoparticles using a microfluidic spray dryer", *Lab Chip* 2011, *11*, 2362. **DOI: 10.1039/c1lc20298g.**

[97] S. Kalepu, V. Nekkanti, "Improved delivery of poorly soluble compounds using nanoparticle technology: A review", *Drug Deliv. and Transl. Res.* 2016, *6*, 319. **DOI: 10.1007/s13346-016-0283-1.**

[98] H. Becker, "Chips, money, industry, education and the killer application", *Lab Chip* 2009, *9*, 1659. **DOI: 10.1039/b909379f.**

[99] C. Holtze, "Large-scale droplet production in microfluidic devices – an industrial perspective", *J. Phys. D: Appl. Phys.* 2013, *46*, 114008. **DOI: 10.1088/0022-3727/46/11/114008.**

[100] T. Femmer, A. Jans, R. Eswein, N. Anwar, M. Moeller, M. Wessling, A. J. C. Kuehne, "High-throughput generation of emulsions and microgels in parallelized microfluidic drop-makers prepared by rapid prototyping", *ACS Appl. Mater. Interfaces* 2015, *7*, 12635. **DOI: 10.1021/acsami.5b03969.**

[101] S.-H. Kim, J. W. Kim, D.-H. Kim, S.-H. Han, D. A. Weitz, "Enhanced-throughput production of polymersomes using a parallelized capillary microfluidic device", *Microfluid. Nanofluid.* 2013, *14*, 509. **DOI: 10.1007/s10404-012-1069-5.**

[102] T. Nisisako, T. Torii, "Microfluidic large-scale integration on a chip for mass production of monodisperse droplets and particles", *Lab Chip* 2008, *8*, 287. **DOI: 10.1039/B713141K.**

[103] T. Nisisako, T. Andoa, T. Hatsuzawa, "High-volume production of single and compound emulsions in a microfluidic parallelization arrangement coupled with coaxial annular world-to-chip interfaces", *Lab Chip* 2012, *12*, 3426. **DOI: 10.1039/C2LC40245A.**

[104] G. Tetradis-Meris, D. Rossetti, C. P. de Torres, R. Cao, G. Lian, R. Janes, "Novel parallel integration of microfluidic device network for emulsion formation", *Ind. Eng. Chem. Res.* 2009, *48*, 8881. **DOI: 10.1021/ie900165b.**

[105] H.-H. Jeong, V. R. Yelleswarapu, S. Yadavali, D. Issadore, D. Lee, "Kilo-scale droplet generation in three-dimensional monolithic elastomer device (3D MED)", *Lab Chip* 2015, *15*, 4387. **DOI: 10.1039/c5lc01025j.**

[106] D. Conchouso, D. Castro, S. A. Khan, I. G. Foulds, "Three-dimensional parallelization of microfluidic droplet generators for a liter per hour volume production of single emulsions", *Lab Chip* 2014, *14*, 3011. **DOI: 10.1039/c4lc00379a.**

[107] M. B. Romanowsky, A. R. Abate, A. Rotem, C. Holtze, D. A. Weitz, "High throughput production of single core double emulsions in a parallelized microfluidic device", *Lab Chip* **2012**, *12*, 802. **DOI: 10.1039/c2lc21033a**.

[108] E. Amstad, M. Chemama, M. Eggersdorfer, L. R. Arriaga, M. P. Brenner, D. A. Weitz, "Robust scalable high throughput production of monodisperse drops", *Lab Chip* **2016**, *16*, 4163. **DOI: 10.1039/C6LC01075J**.

[109] A. Vian, B. Reuse, E. Amstad, "Scalable production of double emulsion drops with thin shells", *Lab Chip* **2018**, *18*, 1936. **DOI: 10.1039/c8lc00282g**.

[110] A. R. Abate, D. A. Weitz, "Syringe-vacuum microfluidics: A portable technique to create monodisperse emulsions", *Biomicrofluidics* 2011, *5*, 014107. **DOI: 10.1063/1.3567093**.

3 Practical realization of microfluidics

Based upon the theoretical fundament of microfluidics in Chapter 1 and its particular utility for forming droplets and materials derived thereof in Chapter 2, one key question remains: where do microfluidic flow cells come from, how are their physicochemical and mechanical properties tailored toward user-specific needs, and how are microfluidic devices handled regarding optimal fluid flow, flow pattern formation, and in key applications such as those in materials design, (bio-)chemical synthesis, and analytics, all with a view to in-flow characterization and manipulation of molecules, particles, cells, and other species?

Owing to the widespread use of microfluidic devices, a myriad of combinations of device designs, building materials, functional elements, and surface modifications exist – each tailored toward a specific application. To systematically introduce the practical realization and handling of these versatile microfluidic systems, in the following sections, we discuss microfluidic device design in a stepwise fashion with increasing complexity. First, we focus on the material basis for microfluidic device fabrication, particularly elastomers and glass. From there, we outline the use of these materials to fabricate microfluidic devices based on elastomer replication with poly(dimethylsiloxane) (PDMS), also termed stamped microfluidics, as well as by glass microcapillary alignment.[1] We then describe some exemplary applications of these devices in the fields of continuous-flow (CF) and segmented-flow (SF) microfluidics, supplemented with info boxes and experimental instructions that may serve as a backup in day-to-day lab work. In SF microfluidics – due to its relevance for materials design (e.g., for polymer particle preparation) as well as in the biosciences (e.g., for encapsulation, cell mimicking, or as bioreactors) – we particularly focus on droplet preparation and their usage. In CF microfluidics, we spotlight on hydrodynamic flow-focusing (HFF), which provides spatiotemporal resolution in the analysis of fast-progressing reactions (e.g., nucleation, self-assembly, or conformational molecular changes) on the millisecond scale, for instance. Before discussing these applications of microfluidic devices in Sections 3.3 and 3.4, we need to focus on their design and fabrication first, which is the topic of

---

1 While these two techniques are well established in the microfluidics community and thus particularly detailed in here, additive manufacturing based on stereolithography (SL), fused deposition modeling (FDM), and inkjet printing has quickly spread among the microfluidics community for flow cell fabrication in the past few years. It is thus only a matter of time until 3D printing will replace conventional flow cell fabrication in terms of resolution, materials availability, surface modification / functionalization, and production throughput. For a detailed view on current additive manufacturing techniques, the reader is referred to key review articles on 3D printing and microfluidics by Folch, Rapp, and others as well as the textbook by Gibson, Rosen, and Stucker focusing on various methods in additive manufacturing, their minimal resolution, and material requirements [1–5].

https://doi.org/10.1515/9783110487701-003

following Section 3.1. From there, we conceptually build up a complete microfluidic experiment made of fluid reservoirs, pumping systems, the as-fabricated flow cell, and optical characterization tools.

Note: To help users new to the field of microfluidics setting up their own flow experiment, it is necessary to specifically mention manufacturers and vendors of materials and equipment. We do this based on our own experience without aiming for completeness or preferring one brand or type of tool over another.

3.1 Fundamentals of microfluidic device design and fabrication

3.1.1 Materials selection guide for microflow cell fabrication

The origins of microfluidics lie in conventional solid-state **electronics**. Indeed, similar fabrication techniques are employed to manufacture both electronic and microfluidic devices, for example, the use of mask aligners for structure generation, which were initially developed for the semiconductor industry. As academic research on microfluidics is primarily devoted toward studying, manipulating, and optimizing design, flow dynamics, and device utilization for individual applications, for example, in specific materials development and (bio-)chemical analyses, microfluidic device preparation requires methods that allow for screening thousands of device parameters in a timely efficient manner. On this account, microfluidic device fabrication based on PDMS is the method of choice, as further outlined in Sections 3.1.2–3.1.6.[2]

Generally, when choosing the material basis for flow cell design, one should not only focus on the suitability of a device material for the microfluidic experiments to be conducted in that device, such as in view of the choice of fluids, solutes, and fluid pressures involved in the experiment, but one should also sketch ahead characterization techniques that are to be used to monitor processes in the flowing fluid streams and compounds formed or transformed in them. These techniques include spectroscopic and scattering methods based on IR, visible light, but also X-ray or synchrotron radiation. For this purpose, it is crucial to choose materials for the device fabrication

---

[2] CF and SF microfluidic experiments can be generally implemented independent of the flow cell's fabrication process (stamped, assembled from glass capillaries, or made by additive manufacturing) or the material basis of it (silicon, glass, polymers, or metals). As this textbook's focus is on the basics of microfluidic device design for (undergraduate) students, we mainly explain theory, experiments, and methods using the examples of stamped flow cells made of polymers – primarily poly(dimethylsiloxane) (PDMS) – and glass microcapillaries. This is both due to the simplicity of such devices as well as due to their wide spread utilization.

that are transparent for the respective radiation. Only then, flow pattern formation, reaction progresses, or other phenomena can be efficiently monitored without loss of information by (microscopic) characterization techniques. For instance, UV-grade quartz glass microscopy slides are optically transparent (transmission > 90%) in a wavelength range of 200–1,100 nm (at 1 mm substrate thickness), standard glass microscopy slides are transparent in a slightly narrower range of 300–1,100 nm, and PDMS is transparent in a range of 240–1,100 nm. Another material, poly(imide) (Kapton®), has a particular strength in the form of transparency at extremely low wavelengths, for example, 99.7% transmission at $\lambda \approx 0.1$ nm; it is therefore particularly suited for X-ray based analysis of structures or phenomena in devices made of that material.

Once these general material considerations for microfluidic device fabrication are made, follow-up questions arise in view of how to process materials such as PDMS, glass, or Kapton® into microfluidic channel systems. This will be in the focus of Section 3.1.4 to Section 3.1.7. Before diving into deeper detail on that, we first consider that for all the material bases just mentioned, there are numerous opinions on the best *environment* for fabricating microfluidic devices, ranging from open lab benches over fume hoods to dust-free flow benches and clean rooms, for example, with ISO 14644-1 standard 5.[3] But what is really required? The following information is based on both authors' opinions and their >10 years of microfluidics experience.

The choice of the right environment for microfluidic device manufacturing is often driven by the user's perception that dust is the "deadly enemy" of every microfluidic experiment, as it can lead to clogging and device failure. This is because dust particles typically have sizes from 0.1 μm to tens of micrometers, which is indeed in the exact size range of microchannels. On this account, microfluidics users often move device fabrication to clean rooms at their location, which are usually operated by semiconductor researchers. In such rooms, they do not only find a nearly dust-free environment, but also a manifold of tools that are required for microfluidic master device fabrication by photolithography (cf. Section 3.1.4), because semiconductor manufacturing and microfluidic device fabrication share the same tools and device basis.[4] However, usage of that equipment is often not for free and must be paid by the hour; even worse, for some users, such advanced clean room facilities are not even available at all. In that case, researchers may consider investing in cleanroom-like working spaces, which ranges from

---

[3] This norm from semiconductor industry requests clean rooms to contain no more than 100,000 particles with size ≥ 0.1 μm, 23,700 particles with size ≥ 0.2 μm, 10,200 particles with size ≥ 0.3 μm, 3,520 particles with size ≥ 0.5 μm, and 832 particles with size ≥ 1.0 μm in air per cubic meter.

[4] This shared list of equipment includes hot plates with homogeneous heat distribution, photoresists, UV mask aligners, and developer baths.

permanently installed facilities (cost: up to several million Euros, depending on the air flow pattern generated inside the cleanroom, for example, roof filter ventilation versus top-to-bottom laminar flow) over walkable cleanroom tents (cost: approx. 20,000–100,000 Euros) to flow benches (cost: approx. 5,000–15,000 Euros) and portable dust-free bench tents (cost: approx. 1,000–5,000 Euros). Yet, these (costly) facilities are actually not necessary. In fact, if a few precautions are taken, **microfluidic devices can also be manufactured at a conventional lab bench** in the "dirty" environment of a chemistry or physics lab, and these devices usually perform just as well as those flow cells fabricated in a dust-free clean room. Focusing on the fabrication of microfluidic devices by combined photo- and soft lithography (PL/SL, Sections 3.1.2–3.1.5), these **precautions include frequent cleaning of any surface**[5] involved in the microflow cell fabrication and frequently checking for contaminations by light microscopy (cf. Section 3.2.4) in the case of transparent samples as well as by confocal microscopy in the case of light-reflecting samples. For surface cleaning, microfluidics users should make use of solvents that do not harm the building materials (e.g., 1-propanol/isopropanol) and that can be fully removed by compressed air (cf. Section 3.1.5).

3.1.2 PDMS-based microfluidics: Devices by computer-aided design

The general **working cycle** for microfluidic device development in its variant of molding flow cells with silicone elastomers is shown in Figure 3.1. First, an idea of a device design is sketched "in silico" by **computer-aided design (CAD)**. That virtual sketch is then plotted onto a transparency that serves as a **photomask in photolithography** to replicate the printed channel structures on a substrate. The resulting relief is then used to mold inverse replicas into a **silicone elastomer based on poly(dimethylsiloxane) (PDMS)**, which is finally equipped with holes for fluid in- and outflow and then bonded to a carrier substrate. One common further step after that process is **microchannel wettability treatment**. The device may then be tested and used, during which ideas for further improvement and optimization will come along, initiating a second iteration of the above circle. In a refined variant of that process, a simulation step is inserted between the idea and the actual device fabrication process to pre-estimate the performance of a device in a virtual rather than a real fashion, as shown in the right sketch of Figure 3.1.

The first production step of the circle of microfluidic device development according to what we have just stated is to come up with a digital image of the

---

5 These surfaces include the (silicon wafer) substrate for microfluidic master device fabrication as well as any PDMS slab and glass substrate. By the way, as we are talking about avoiding dust contaminations in microfluidic devices, note that stamped flow cells should be covered with easy-to-remove tape, e.g., Scotch Tape, 3M, at the inflow and outflow ports for storage.

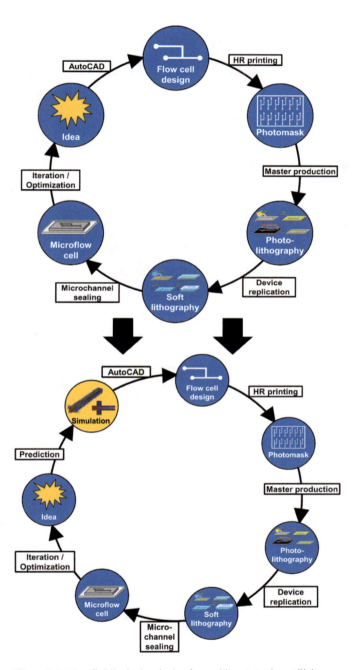

Figure 3.1: Microfluidic device design by rapid prototyping utilizing a combination of photo- and soft lithography (top) as well as refined device design introducing fluid-flow simulations as an additional step for optimizing the fluid flow and microchannel design and thereby ideally reducing the number of experimental iterations to identify the optimal device design (bottom). Adapted from Y. Xia, G. M. Whitesides, *Angew. Chem. Int. Ed.* 1998, *3*, 550. **DOI: 10.1002/(SICI)1521-3773 (19980316)37:5<550::AID-ANIE550>3.0.CO;2-G**; copyright 1998 WILEY.

Treating microfluidic devices like electronic circuits (Part 1)

Counterparts of many active components of electronic devices can be found in microfluidic devices.

⇩

Electronic resistors correspond to fluid-filled microchannels, transistors to elastomeric valves, and capacitors are analogous to microchambers with membranes that expand or contract.

Many analogies exist among equations to describe flow phenomena in microfluidics and electronics. Assuming steady flow without energy dissipation at low Reynolds number inside a microfluidic device, fluid flow is described by the linear Stokes equation, and the fluidic resistance is given by the **Hagen–Poiseuille law** (introduced in Chapter 1.1.2). This equation is analogous to **Ohm's law** for a DC circuit, where a potential drop ΔU is described over the length of a wire with resistance R that is fed by an electrical current I.

$$\Delta p = R_{hydr} Q \quad \Longleftrightarrow \quad \Delta U = R I$$

Thus, **voltage U**, which is the energy per charge, **corresponds to pressure p**, which is the energy per volume, and electrical current (electrical charge per time) I is analogous to the volumetric flow rate (volume per time) Q.

 Treating microfluidic devices like electronic circuits (Part 2)

Based on the analogies between the Hagen–Poiseuille law (introduced in Chapter 1.1.2) and Ohm's law, it is adequate to **approximate flow rates in a microfluidic device by a network of microfluidic channels rather than a detailed view on the velocity field and volumetric flow rate Q**. In good approximation, the **hydraulic resistance R_{hydr} obeys the same rules for series and parallel connections as the electric resistance R** in the linear circuit theory. For two hydraulic resistances R_1 and R_2, it follows:

$$R_{hydr}(\text{serial}) = R_1 + R_2 \qquad R_{hydr}(\text{parallel}) = \left(\frac{1}{R_1} + \frac{1}{R_2}\right)^{-1}$$

Accordingly, we can apply **Kirchhoff's first and second circuit laws to describe fluid flow in microchannel networks**:

At any junction in a circuit, the sum of current arriving equals the sum of current leaving the junction (→ **charge conservation**).	Around any closed loop within a circuit, the sum of potential differences across all elements is zero (→ **energy conservation**).
⇩	⇩
The sum of fluid flow entering and leaving any junction in the circuit is zero.	The sum of all pressure differences within a closed loop of the microchannel circuit is zero.

microchannel network. In simulation-based rapid prototyping, as shown in the right sketch of Figure 3.1, simulation tools themselves often provide a simple drawing and design tool that can be utilized to sketch microchannels. However, for drafting complex devices with functional units, simple drawing programs may not be sufficient. Moreover, one has to keep in mind that we generate an image of our microchannels not just simply for visualization purposes, but it will serve as a digital basis for fabricating corresponding photomasks, which contain a to-scale image of the microfluidic device that is then transferred to a substrate employing the actual photolithography process (cf. Section 3.1.4). Thus, we need to choose drawing programs that handle file types compatible with the process of photomask generation (cf. below). As the development of microfluidic devices has many of its origins in microelectromechanical systems (MEMS), where CAD tools are commonly employed for circuit board design, they are likewise widely employed in microchannel network design in PDMS-based microfluidics (cf. info boxes below).

The list of these CAD design tools is long and includes programs such as AutoCAD, Solid Works, CleWin, and LinkCAD. Focusing on AutoCAD as a commercial,[6] yet widely distributed CAD software, the following sticky notes compile frequently used commands in 2D microchannel network design.

AutoCAD commands for microflow cell design (Part 1)

Hints:
- Use PC's hardware acceleration
- Cancel selection of objects by ESC
- Repeat command by SPACE
- Mouse selection: left-to-right = only objects inside box; right-to-left = everything touching box
- Save drawings as *.dwg or *.dxf depending on print service or printer

Commands:

OSNAP	Set running object snap mode; define preferences (ENDpoint, MIDpoint, CENter, INTersection, EXTension, PARallel)
LINE	Draw a line (enter length after defining first point by clicking on position in space or enter value)

---

[6] AutoCAD poses a significant investment (approx. 2,000 Euros for a one-year subscription). So, when setting up the whole process line of microfluidic device fabrication by combined PL and SL, the reader may consider either to choose a free tool instead or to check whether he/she is eligible to subscribe for a free student version of AutoCAD.

AutoCAD commands for microflow cell design (Part 2)

Commands:

CIRCLE — Draw a circle (enter radius after defining first point by clicking on point in space)

RECTANGLE — Draw a box (enter parameters a and b by typing "D" after defining first point in space)

MOVE — Select an object by clicking on reference point (e.g., corner or midpoint), then drag and click to place object

TRIM — Shorten object by deleting overlapping structure (select structure for defining cutting line, then SPACE, then object that needs to be cut, then ENTER)

AutoCAD commands for microflow cell design (Part 3)

Commands:

BLOCK — Associate objects to block (name structure, then add to library for reuse)

EXPLODE — Break a composite object (e.g., blocks, polylines) into its components for modifying its parts individually

ARRAY — Create arrangement of objects, e.g., 2D rectangular pattern (select object as basis, then define number of rows / columns, forming - by definition - up / right)

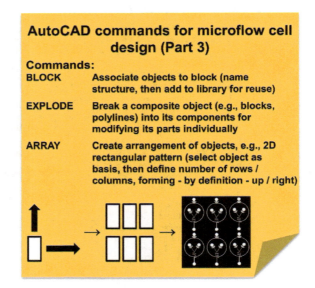

> **AutoCAD commands for microflow cell design (Part 4)**
>
> **Commands:**
>
> EXTEND — Extend object to join edges of neighboring structures (define where to extend to, select object for extension)
>
> MIRROR — Duplicate object (select object, then SPACE, select two points along pre-defined reflection axis, choose YES / NO for deleting / keeping original object)
>
> OFFSET — Relocate object (select object, specify offset distance or define two points as offset distance)
>
> ERASE — Delete components of object (select corresponding single / multiple parts)
>
> Command aliases (C, then ENTER, instead of CIRCLE, then ENTER) can be found in the files acad.pgp or acadlt.pgp.

With a CAD file of the desired microchannel network at hand, the next step is to choose the material basis for the photomask and match the minimal feature size in the CAD file with the resolution offered by the individual type of photomask. Here, we can distinguish three types of masks that can be employed in microfluidic device fabrication by photolithography: quartz glass, soda lime, and polymer foils or transparency plastics. Despite a trade-off in resolution and stability compared to chromium masks on soda lime or quartz glass, **flexible photomasks mostly made from polyester** are easy to handle, and can be made at any size for comparably low cost. They are thus most suited for beginners in microfluidics who may face a long list of initial investments and need to plan carefully how to spend their money. Moreover, low-cost polymer foils are always useful in the process of flow cell design optimization. Then, a DIN A4 page corresponding to a size of 10″ × 12″ will be an investment of just approx. 150 Euros when printed at 128,000 dpi resolution. Assuming a substrate diameter of 2 inches (= 50.8 mm) like a silicon wafer, we can fit approx. 25–30 different master device designs on a single DIN A4 page or photomask. In addition, on each individual 2″ spot, we can usually fit two or more rows of flow cells designs, as shown in Figure 3.2, thus allowing for inclusion of a large number of different design variants on a DIN A4 photomask. Due to their significant costs, one may choose more sophisticated glass masks at a stage where an optimal microchannel design has been identified and shall be printed with higher resolution for improved edge sharpness, for instance. Along these lines, a key benefit of glass or soda lime masks is the minimal feature size, which is usually in the nanometer range, whereas for polymer foils, the minimal resolution is rather 2–15 μm. Another benefit of glass masks is that

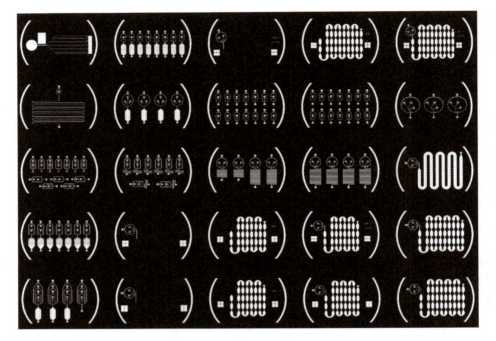

Figure 3.2: Photomasks for fabricating microchannels on receiver substrates by photolithography for subsequent replication by soft lithography. The image shows a 5 x 5 array of different business-card-sized photomasks that can be cut from the convolute page and used in further hard-lithography working steps to produce microfluidic devices from them, as outlined in the following subchapters. In these photomasks, all areas are black except for the regions that are to become the channels. These regions are transparent to allow UV light to pass through the mask to expose the photoresist underneath on the substrate surface.

they are solid and inflexible, which makes them easier to handle on UV-exposure setups, where the masks are mounted to. Here, polymer foil-based photomasks require a solid, UV-transparent support to which masks can be attached to. As this is usually achieved by sticky tape, we introduce an additional spacer between the photosensitive substrate and the photomask, which may cause additional UV light scattering and reduced resolution.

Independent of the type of photomasks, these are often mounted onto a so-called **mask aligner** for transferring the microchannel image onto a UV-sensitive substrate. If the experimenter does not use glass photomasks, which usually contain just one device design, but a polymer foil with an array of designs, the photomask needs to be carefully cut into smaller, mask aligner-compatible pieces. For that, the user should use non-powdered gloves (to avoid making scratches or leaving fingerprints on the masks that may scatter the UV light) and a sharp scalpel or razorblade for the cutting.[7]

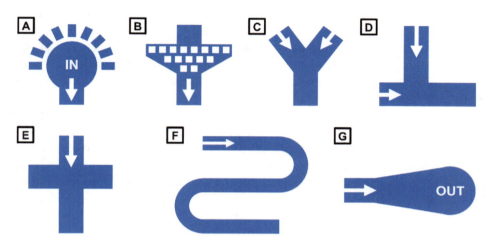

Figure 3.3: Common device design features in stamped microfluidics. (A) Inflow port, (B) filters, (C–E) Y-shaped, T-shaped, and X-shaped microchannel junctions, (F) meander-shaped microchannel delay lines, (G) outflow port.

3.1.3 PDMS-based microfluidics: Basic channel-system elements

Having discussed the software tools for photomask design as a basis for microfluidic master fabrication by photolithography, we focus next on some basic design features found in microflow cells and their role in flow pattern formation.

In general, each microchannel system consists of a couple of fluid inlets, the channels themselves, and one or more fluid outlets. In most cases treated in this book, the heart of the channel system features one or more channel *junctions*, which may have either T-shaped, Y-shaped, or X-shaped geometries. A further typical element is delay lines such as meanders. On top of that, the authors' experience has shown that in some cases, it is beneficial to include filter units as further elements, positioned right behind the fluid inlets. Figure 3.3 compiles schematics of all these typical microfluidic channel elements, which we now discuss one by the other.

In regard to **inlets**, there is not much to discuss; they are usually patterned as a simple circular marker of about 1 mm diameter that will later be punched through

---

[7] Smaller photomask pieces are usually similar in size to business cards. A folder for business card storage is thus recommended to organize and protect the photomasks, which can be generally reused a thousand times if stored and handled properly.

with a biopsy needle to allow tubing to be plugged into the device for fluid injection. In that step of the device fabrication process, it is helpful to guide the user's eye for accurate manual punching. For this purpose, it is practical to pattern a semi-circular corona of lines around the inlet circle, as seen in Figure 3.3A.

Right behind the inlet, it is often helpful to have a **filter unit** composed of a two- or three-row array of diamond or square-shaped pillars, as shown in Figure 3.3B. These pillars block debris in the to-be-injected solutions to enter the downstream channel system, where such debris might lead to clogging or uncontrolled flow conditions.

Behind the inlet and filter units, the actual channel system is placed. Typical elements used in many microfluidic experiments, such as those discussed in further detail in the following subchapters, are T-shaped, Y-shaped, or X-shaped **channel junctions**, as shown in Figure 3.3C–E. In those, different fluid streams can be brought into contact to either flow alongside each other in the variant of CF microfluidics or to form emulsion droplets or plugs in the variant of SF microfluidics. If the resulting multicomponent fluid stream, be it a continuous or a segmented one, shall then be further extended before exiting the device, for example, to maintain the well-arranged mutual arrangement and spacing of the different fluid components for a while, a **delay channel** in the form of a **meander** is an element of choice, as shown in Figure 3.3F. Eventually, though, the fluid stream must and will exit the device though an **outlet**, which is geometrically equal to the inlets discussed above, as shown in Figure 3.3G.

As simple as the just-mentioned fundamental channel-system elements appear, there are some nitty-gritty details to be known about them and about the microchannel design in general to optimize the performance of a microfluidic device. In the following, we first discuss some of such general details that apply to the whole microfluidic system, before we also discuss some details about the single-channel elements treated above, in the same order as they were introduced in the preceding paragraph.

The first fundamental design feature of every patterned microfluidic device is its **aspect ratio**, h/w. Tailoring that ratio has been previously discussed as a straightforward method to minimize surficial contact between flow-focused fluid jets and the upper and lower microchannel walls, thereby avoiding wetting (cf. the end of Section 2.3.3). However, beyond the shape of a flow pattern, the microchannel aspect ratio also has a significant effect on the velocity profile within a microchannel and thus the distribution of diffusing reagents or analytes in a channel. For example, velocity profiles inside microchannels with an aspect ratio of 1:1 and 1:2 are usually parabolic in both height and width, as shown in Figure 3.4A. Here, it is challenging to predict the residence time for a molecule or particle as it apparently depends on its spatial location within the liquid flow. For instance, diffusion that occurs transversal to the flow across the microchannel width may result in a significant increase in the residence time of compounds of interest when moving away from the center of the liquid flow toward the microchannel walls. Due to the resulting distribution of residence times, analysis of time-dependent events that occur within the channel such as nucleation, self-assembly, or chemical reactions become difficult to predict and quantify. By contrast, at a higher aspect ratio such as 4:1 and particularly 8:1, the characteristic parabolic flow profile for pressure-driven flow in microchannels both along the top-to-bottom (height) and the left-to-right (width) microchannel centerline vanishes. Instead, a nearly homogeneous flow front develops across the width of the microchannel (Figure 3.4B, bottom panel), which exhibits nonuniformity solely close to the walls, while only across the height of the microchannel, the typical parabolic flow profile for pressure-driven flow remains. Such type of velocity profile allows lateral diffusion to occur across the microchannel width without accompanying changes in the residence time of the diffusing molecules. Therefore, microflow cells with high aspect ratio are often more desirable for quantitative studies as well as homogeneous nucleation and self-assembly processes (cf. Section 3.3.2) than microchannels with a square cross-section.

The second fundamental design feature of microfluidic devices is the way how channel-direction changes are designed. For the ease of device design in CAD software, the inflow channels are usually purely rectangular, as sketched in the left of Figure 3.5.

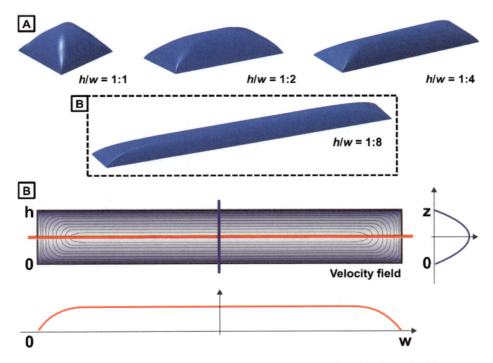

Figure 3.4: 3D profiles of velocity fields for Poiseuille flow in rectangular microchannels with different aspect ratios. (A) 3D depiction of laminar flow profiles in microchannels with aspect ratios of $h/w = 1$, $h/w = 0.5$, $h/w = 0.25$, and $h/w = 0.125$. (B) Contour lines for a velocity field with line scans in blue and red (left), indicating the typical parabolic profile lateral to the flow direction (right). Adapted from H. Bruus, *Lab Chip* 2011, *11*, 3742. **DOI: 10.1039/c1lc20658c;** copyright 2011 Royal Society of Chemistry.

Yet, rectangular microchannel features pose the risk of facilitating the formation of **fluid-flow dead zones,** as also sketched in the left of Figure 3.5. In such dead zones, especially particulate fluid flow (e.g., containing cells or microparticles) is prone to cause agglomeration and even clogging of a rectangular microchannel, as in this dead zone fluid elements are not, or at least not sufficiently, exchanged over time. To circumvent this drawback, it is recommendable to use curved channels, as sketched in the right of Figure 3.5. Note: the same device design considerations also hold for outflow channels, especially in CF microfluidics. For instance, meandering outflow channels, which are commonly utilized as delay lines to achieve diffusion-based mixing of co-flowing fluids, should likewise be designed to avoid the afore-discussed dead zones in any meander turn, as these may lead to partial or even full blocking of the microchannel cross-section due to material agglomeration in flow dead zones, thus entailing high backpressure at a high number of microchannel turns, or even causing clogging.

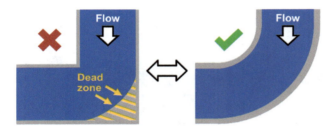

Figure 3.5: Microchannel design for optimizing the flow in microfluidic devices. For beginners as well as users who aim for time-efficient drawing by CAD, it is handy to draw microchannels as rectangular networks. In corresponding flow cells, however, this design may cause dead zones, which are prevented by adapting or smoothening the microchannel design.

A row of further comments must be made on the single device elements introduced in Figure 3.3. In regard of filter units, it is to state that those can lead to unwanted and uncontrolled emulsion droplet formation in SF microfluidics when there is **backflow** in the device, which particularly occurs during the period of the microfluidic device start up.[8] During such backflow, geometric features like filter pillars act as obstacles at which the continuous flow of the two injected phases is dispersed, as shown in Figure 3.6A. As this accidentally formed emulsion is later (once the device operates in the correct direction after reversion of the backflow) re-ejected through the droplet-forming nozzle, the desired constant inflow of liquids is disturbed, leading to uncontrolled flow conditions and thus inhomogeneous material properties of the microfluidic product (e.g., droplet size polydispersity). Hence, in this variant of

---

8 How to damp fluid backflow? – Check Figure 3.16.

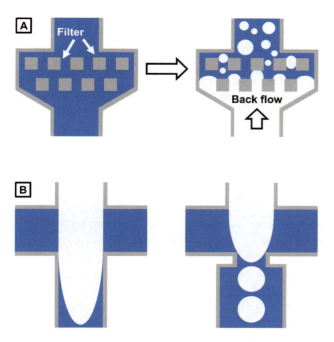

Figure 3.6: Microchannel design for optimizing the flow in microfluidic devices. (A) Arrays of squared or diamond-shaped posts within a flown-through microchannel serve as barriers that filter debris. They, however, can cause accidental emulsion-droplet formation in the case of backflow during the starting period of a SF microfluidic experiment. (B) Formation of emulsion droplets from complex liquids (e.g., shear thinning or low-surface-tension liquids) may require support by geometric features at the droplet-forming cross-junction that enhance local shear forces and promote droplet breakup at a defined position within the microfluidic device.

microfluidics, it may be beneficial to NOT use filter units in the microfluidic device, but instead, rather carefully filter the to-be-injected fluids beforehand.

In contrast to this potential drawback of filters for controlled microfluidic emulsification, in regard of the actual drop-making elements, that is, the channel cross-junctions, there's some geometric features that can be extra beneficial to promote controlled droplet breakup. For instance, instead of designing a microchannel cross-junction with uniform microchannel aspect ratio, as shown in Figure 3.3C–E, experimenters may introduce a narrow microchannel section directly behind the cross-junction to increase the shear forces on the to-be-dispersed phase jet to induce droplet breakup, as sketched in Figure 3.6B.

Whatever type of fluid flow is formed in a microfluidic channel system must exit it at some point. This can occur via two ways. In one possible realization, the fluid flow can exit the microfluidic device perpendicular to the microchannel network via tubing. In another possible realization, the fluid flow can exit the device along the continuous flow and thus in plane with the microfluidic experiment. The

Table 3.1: Diffusion times depending on the travel length as found in microflow cells.

Distance (μm)	Diffusion time (s)			
	Heat	Molecule	Protein	Cell
1	0.0001	0.001	0.01	10
10	0.01	0.1	1	1,000
100	1	10	100	100,000
1,000	100	1,000	10,000	10,000,000

latter device design is frequently utilized in microfluidic fiber spinning by polymer extrusion, for instance. Here, to prevent clogging of the flow cell due to a change in the flow direction of the as-fabricated polymer fiber, as it would occur in conventional flow cells with vertical fluid outflow ports, the experimenter may utilize flow cells with fluid outflow ports in the flow direction. As a side effect, the undisturbed flow of a continuous jet (or as-formed polymer fiber) enables microfluidics users to build up continuous extensional forces on the fluid jet to align fibrils in the flow direction for optimal fiber mechanical properties. Freestanding jets that exit a microfluidic device are likewise of interest for characterization purposes by harsh illumination techniques involving X-rays or synchrotron radiation [6]. Here, continuous jets that extend outside the microfluidic device in air, which are made of two liquid streams, allow for analyzing reaction progress at liquid–liquid interfaces induced by diffusive mixing (cf. Section 3.3) without the constraints of microchannels (scattering effects) and device building materials (absorption).

To close our discussion on microfluidic channel elements, we spend some few more words on those elements that serve for the most common function in microfluidics, which is to bring fluid streams into controlled contact, thereby inducing a most fundamental operation: mixing.[9] Mixing by diffusion is generally perceived as a rather slow process on macroscopic scales, as shown in Section 1.3.5 (also cf. Table 3.1). In microfluidics, by contrast, due to the short lengthscales involved, even that slow process has a marked effect (as also shown in Section 1.3.5). This is favorable for processes where mixing controls the kinetics, such as nucleic-acid

---

[9] Of course, mixing is not the only function microfluidic devices can perform. Further functional device elements include basins for liquid storage, electrodes to build up electric fields, e.g., for fluid flow manipulation, and valves to controllably seal and open microchannels, for instance. To not blow up the frame of this text book, we focus on mixing, as this function can be easily managed by new microfluidics users, and we can utilize this function in both materials science (synthesis), the bio sciences (resolving folding kinetics and binding events, among others), and analytics (e.g., biochemical assays).

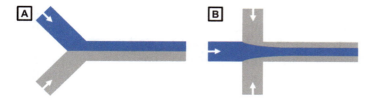

Figure 3.7: Different variants of fluid-stream contact in continuous-flow (CF) microfluidics. (A) Y-shaped microchannel junction for injecting two fluids. At low Reynolds-number flow, these fluids steadily mix by diffusion at their interface along the outflow channel. (B) Microfluidic flow-focusing junctions for forming a narrow liquid jet from three inflows, whose width is controlled by the flow rate of the upper and lower inflow channel.

arrangement and protein folding in biology as well as nucleation processes in materials chemistry [7–9]. Effective mixing can also push back the formation of side products in chemical reactions such as in the initiation of polymerization processes by fast addition of an initiator, thereby yielding polymers with narrow chain-length distribution. In CF microfluidics, **hydrodynamic flow-focusing** can be perceived as the most basic mixing experiment [10]. It usually involves two miscible liquids, one of them being narrowed into a thin liquid jet by a second perpendicular inflow from either one side in a T-shaped or a Y-shaped junction, as shown in Figure 3.7A, or from both sides at a microchannel cross-junction, as shown in Figure 3.7B. This leads to well-defined stationary concentration profiles, which can be mapped within a flow cell such that every point along the liquid jet refers to defined reaction time and intermediate (cf. Section 3.3.2). The time it takes to achieve quantitative mixing of the co-flowing liquids is controlled by the jet's diameter, which is set by the **flow-rate ratio** of the inner and outer phase. This way, liquid jets with diameters as small as tens of nanometers can be generated, corresponding to microsecond mixing times, which are hardly achievable in bulk mixers.

Exemplarily, in a CF experiment, two reagents A and B are injected into a microfluidic device. Owing to the prevalence of laminar flow conditions in microfluidics, molecular diffusion governs the mixing of these fluid streams at their liquid–liquid boundary as they flow downstream the microchannel cross. If we assume highest possible conversion of the reagents A and B toward a product C when all reaction partners are homogeneously mixed, the microchannel layout needs to be adapted for a specific set of flow rates to assure that only the desired product exits the microfluidic flow cell. Only then, the reaction takes place within the highly defined environment of the microfluidic flow cell. Thus, the performance of a microfluidic mixer strongly depends on a careful design of the corresponding microchannel network.

For fast microfluidic flows or fast reaction kinetics, however, diffusion is not sufficient to fully mix fluids before these undergo a reaction or exit the flow cell. In this situation, rapid mixing with low reagent consumption can be achieved using **geometric mixers**, as shown in Figure 3.8. In a first approach to improve microfluidic mixing between two compounds, one can simply increase the channel length and thus the time of fluid contact to allow them to fully mix. One may also decrease the overall flow rates, thereby extending the residence time. As a further means, one may shorten the distance over which fluids must travel and mix by diffusion, for example, by increasing the interfacial area between them over which diffusion occurs. If the footprint of a microfluidic device needs to be small, such as in advanced system integration for handheld end-consumer or in-field devices, however, the latter three means cannot be realized due to lack of space on the microfluidic chip. In that case, one has to redesign microchannel networks that allow for efficient mixing within a short flow microchannel length and small flow cell area, respectively, namely **micromixers** [11–13]. Among the multitude of components that can be integrated in microfluidic devices (pumps, sensors, filters, valves, and mixers), micromixers address the demand to efficiently homogenize reaction mixtures. There are two different types of micromixers: *passive* mixers and *active* mixers. The prior ones rely on the microchannel geometry and either increase the area over which diffusion occurs or induce chaotic advection, thereby decreasing the diffusion path and mixing times. The latter ones rely on the input of additional energy and utilization of external sources, which includes stirring, ultrasound, pressure, temperature, and electric as well as magnetic fields.

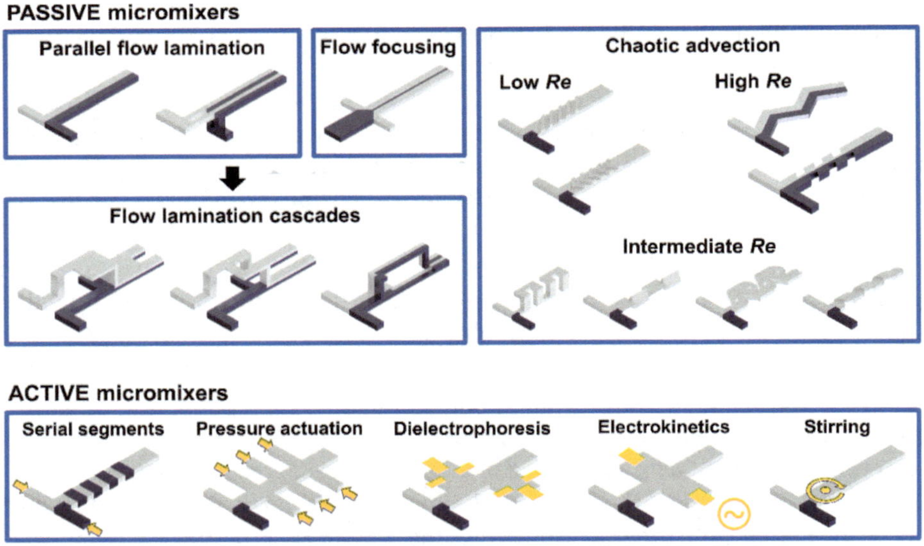

Figure 3.8: Different types of passive and active microfluidic mixers applied for homogenization of two fluids (gray and blue). Upper row: passive micromixers serving to increase the interface between two fluids by flow lamination or chaotic advection, the latter one grouped into mixer geometries for optimized fluid flow at low, intermediate, and high *Re* numbers. Lower row: active micromixers integrating pressure, electric fields, acoustics, and stirring in the mixing process.

For passive mixers, it is advantageous that these do not require external resources – except for pumps required for the fluid transport – compared to active mixers, where external acoustic and voltage generators or pressure manipulates the fluid flow. For instance, to decrease the footprint of the microfluidic mixer in the microchannel network and to improve mixing times at the same time, two fluid streams (as injected into a Y-shaped junction) need to be folded and/or split into numerous substreams that are later recombined (cf. Figure 3.8) [14]. If a laminar flow of two miscible phases is split and rejoined multiple times to yield a parallel stream of fluids, one obtains 2^N fluid layers after N splitting and joining elements, respectively. When building up cascades of these laminar mixers consisting of N elements, the overall mixing time reduces by a factor of 4^{N-1}.

Controlling flow patterns on the micrometer scale is not only crucial to achieve efficient and fast microfluidic mixing (Figure 3.9A), but also vital in another key application of microfluidics, which is extraction from and filtering of femto- to microliter samples. For that purpose, the specific type of microflow cell design contains four microchannels that coincide in a straight microchannel – two from each side – commonly named **H-filter** (referring to the microchannel design in Figure 3.9B). This type of flow cell can be utilized to extract small colloidal or even molecular

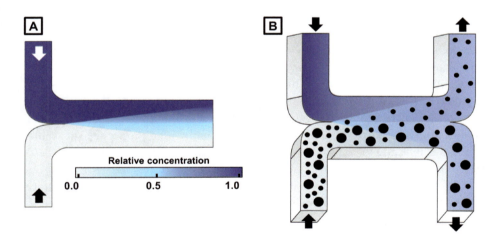

Figure 3.9: Separation of two species of small analytes and subsequent extraction of the species with the higher diffusion coefficient in an H-filter. (A) Conventional microchannel T-junction for mixing two species by transversal diffusion at low Re number flow. (B) The same concept can be utilized for separation of two species of small analytes and subsequent extraction of the species with the higher diffusion coefficient. In an H-filter, mixing is improved by employing narrow, long microchannels in which the concentration gradient will be large in the vertical direction, and diffusion across the fluid boundary can take place on extended timescales. In addition, to-be-filtered species injected into the H-filter should exhibit large diffusion coefficients such that their diffusive flux is fast.

analytes from complex liquid samples, for example, biological colloidal dispersions such as cells, bacteria, and viruses, but also nanoparticles and dust in a continuous fashion. Compared to conventional (macroscopic) extraction, purification, and replacement techniques, an H-shaped microflow cell does not require a membrane or filter unit incorporated into the microflow cell. Therefore, an inherently challenging task in microfluidic device preparation, which would be the combination of a device material such as PDMS with a membrane or filter material without leakage and with micrometer-scale precision, is circumvented.

With its specific design, the H-filter allows for separating one differently mobile species from a mixture of two of more species, yielding a purified solution of the faster-diffusing species during *in situ* extraction. As Reynolds numbers are generally well below one in microfluidic channels, no convective mixing of fluids occurs. Thus, the only means by which solvents, solutes, and suspended analytes move in a direction transverse to the direction of flow is by diffusion, as introduced in Section 1.5 and just discussed again above. For that, the complex mixture is injected into the H-filter from one side, and a collection buffer into which the faster-diffusing species will migrate into, is injected from the other side. The efficiency of the extraction process now depends on the diffusion coefficient of each species, the contact time of the co-flowing fluid flows, and the dimensions of the common microchannel. As the diffusion coefficient scales with the inverse of the analyte size (Stokes–Einstein eq. 1.47), differences in the diffusion coefficients can be used to separate molecules or particles over time. Small molecules have large diffusion coefficients and will therefore move a longer average distance per time than large molecules that have small diffusion coefficients. As such, whereas the small molecules of the entering to-be-filtered stream make it into the co-flowing host stream, the large molecules do not make it and remain in the original stream, as sketched in Figure 3.9. The time spent flowing in the common diffusion area is proportional to the length of the microchannel, so proper design of that channel is required to control the extraction of analytes with high diffusion coefficients from a sample also containing species with lower diffusion coefficients. The H-filter can also be operated in such a way to allow gravitational or electrostatic fields to separate heterogeneous samples into individual fractions, as commonly employed in Field Flow Fractionation (FFF).

3.1.4 PDMS-Based microfluidics: Preparation of microfluidic master devices by photolithography

After digitally sketching a microfluidic device, the most common technique for transferring its microchannel design into the "real world" is based on a flat substrate onto which the microchannel structure is patterned such to be replicated by elastomer molding in a later, separate step (cf. Chapter 3.1.5). The fabrication of this

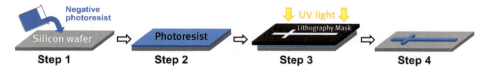

Figure 3.10: Manufacturing of a microchannel master structure by photolithography (PL). In Step 1, a small portion of a photoresist (here, a negative photoresist from the SU-8 series offered by MicroChem) is poured onto flat, smooth substrate (e.g., a polished silicon wafer) and then, in Step 2, it is processed into a defined layer thickness by spin coating. In Step 3, the coated substrate is exposed to UV light through a photomask projected in a computer-aided design (CAD) program, for example, AutoCAD®. In Step 4, the imparted microchannel structure is worked-out by subsequent thermal hardening followed by removal of unhardened areas in a so-called developer bath, usually based on propylene glycol monomethyl ether acetate (PGMEA), for example, mr-DEV 600 (MicroChem). As a result, a positive relief of the microchannel design from the photomask is obtained that may then be subject to negative replication in an elastomer such as poly (dimethylsiloxane) (PDMS), as detailed in Section 3.1.5.

microstructured substrate – commonly named *microfluidic master device* – is generally based on **photolithography** (PL), as illustrated in Figure 3.10.[10] This process starts from a flat clean substrate – usually a polished disk-shaped silicon wafer – coated with a thin layer of a **photoresist**, normally SU-8.

The thickness of the photoresist layer defines the height of the later microchannels; hence, it is crucial to apply layers of defined thickness to the substrate. For that purpose, we pour a coin-sized blob of a photoresist resin onto the center of the wafer and distribute it evenly on the surface by spinning it at a high speed. Typical process parameters for different SU-8 resins and layer thicknesses are provided in the info boxes below.

Depending on the viscosity of the photoresist, the spinning rate, and the process time, the thickness of the coating can be controlled with micrometer precision. The coated wafer is then heated to evaporate solvent and to pre-solidify the coating, usually first at 60 °C and then also at 95 °C, in a process step named **soft bake**, and then cooled again. Depending on the type of resin used and depending on the thickness of it on the substrate, different heating times must be used at the two heating temperatures. After the heating, again depending on the thickness of the photoresist, the experimenter may have to include a controlled gentle cool-down procedure – for example, from 95 °C over 65 °C back to room temperature instead of directly taking the coated substrate from the hotplate. This way, we avoid material stress

---

[10] With the advent of high-resolution additive manufacturing (AM), photolithography can soon be replaced by the fabrication of microfluidic master devices for further elastomer replication via micro-stereolihography (μSL) in a single step. While the resolution of commercial μSL is not yet on the exact same level as that of PL, microchannel feature sizes on a flat substrate as small as 20 μm can be achieved already.

Photolithography recipes for negative photoresists (NANO™ SU-8 series) Part 1

The following tables summarize the **key process parameters of master device fabrication** utilizing photolithography. Depending on the desired microchannel height, different photoresist formulations are applied, each designed for a specific working range and layer thicknesses, respectively.

10 μm ↑	SU-8 2007	Speed (rpm)	T (°C)	t (s)
	Spin coating	a) 500 b) 1,800		a) 12 b) 40
	Soft bake		a) 65 b) 95	a) 60 b) 150
	UV exposure			23 ON, 40 OFF, 23 ON
	Post-exp. bake		a) 65 b) 95	a) 60 b) 180
	Developer bath			180
	Hard bake		170	60

25 μm ↑	SU-8 2025	Speed (rpm)	T (°C)	t (s)
	Spin coating	a) 500 b) 3,600		a) 12 b) 40
	Soft bake		a) 65 b) 95	a) 60 b) 240
	UV exposure			23 ON, 40 OFF, 23 ON
	Post-exp. bake		a) 65 b) 95	a) 60 b) 180
	Developer bath			180
	Hard bake		170	90

Photolithography recipes for negative photoresists (NANO™ SU-8 series) Part 2

50 μm ↑	SU-8 2025	Speed (rpm)	T (°C)	t (s)
	Spin coating	a) 500 b) 1,400		a) 12 b) 40
	Soft bake		a) 65 b) 95	a) 120 b) 840
	UV exposure			38 ON, 40 OFF, 38 ON
	Post-exp. bake		a) 65 b) 95	a) 60 b) 420
	Developer bath			300
	Hard bake		170	90

100 μm ↑	SU-8 100	Speed (rpm)	T (°C)	t (s)
	Spin coating	a) 500 b) 3,100		a) 12 b) 40
	Soft bake		a) 65 b) 95	a) 600 b) 1,920
	UV exposure			42 ON, 40 OFF, 42 ON
	Post-exp. bake		a) 65 b) 95	a) 60 b) 600
	Developer bath			720
	Hard bake		170	120

due to potentially inhomogeneous cooling within the photoresist, which could lead to delamination of the photoresist from the substrate as well as to undesired cracks that may scatter the UV-light in the subsequent patterning step, thereby reducing the microchannel resolution. Note: at this point, the coating is solid but fragile and can easily (which means: accidently) be removed with a solvent. Further note: in some cases, the photoresist may not exhibit the desired uniform height profile throughout the substrate surface, but be more similar to a donut shape that is higher at the rim of the disk-shaped substrate than in the middle. Usually, this effect can be pushed back by reduction of the initial amount of photoresist that is poured onto the substrate before spin coating. In a less elegant fashion – for thin layers of photoresist with a donut-shaped distribution profile on the substrate – the experimenter may also carefully squeeze the soft photoresist during the forthcoming UV-exposure step by gently pressing the photomask onto the photoresist. If a mask aligner (e.g., MJB3 or MJB4, SÜSS MicroTec SE; EVG610, EV Group E. Thallner GmbH; Model 200, SPS-Europe B.V.) is employed to achieve the forthcoming UV patterning, the stage carrying the photo resist-coated substrate may also be pushed stronger to the mask in the contact mode of the apparatus to level out the inhomogeneous distribution of photoresist on the substrate.

In the next step, the photomask is placed on top of the coated wafer, and the two are exposed to ultraviolet (UV) light for a time depending on the type and thickness of the photoresin layer, as compiled in the info boxes above. The light passes through the transparent regions of the photomask but is blocked by the black parts. The photons that penetrate create a Lewis acid that then triggers **solidification of the photoresist** by crosslinkage of its epoxy-functional polymers, as shown in Figure 3.11. To further fuel the crosslinking reaction, the wafer is heated for a few minutes right after the UV illumination in a step named "post-exposure bake", again at two different temperatures at different times each, once again depending on the layer type and thickness (see info boxes for details). During that, the crosslinking of the resin happens in the zones that were illuminated before, whereas the unexposed regions stay un-crosslinked. Some care is adequate during that step: while it has to be long enough to achieve good crosslinking, it shall also not be too long to avoid overreaction into the unexposed zones of the coating. Adequate curing times are given in the info boxes above. After that step, the wafer is placed in a solvent wash named **developer bath**, usually based on propylene glycol monomethyl ether acetate (PGMEA), for a period of time (again, see info boxes for details) to remove the unexposed and un-crosslinked portions of the photoresist. This leaves behind the crosslinked parts, which form a relief-shaped copy of the photomask in the form of "positive" (i.e., protruding) features where the light has exposed the resin. In a final step, the so-fabricated master device is subject to another heating procedure (named **hard bake**; again, see info boxes for details) to finalize the resin crosslinking.

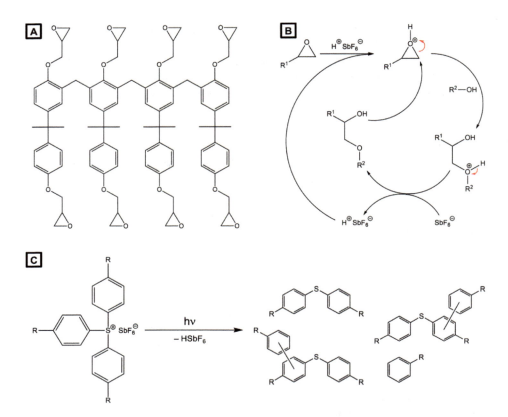

Figure 3.11: Reaction mechanism of SU-8 photo-crosslinking. (A) Chemical structure of Shell's EPON SU-8. (B) Crosslinking of the SU-8 by epoxide ring-opening according to Louis J. Guerin. (C) Photo-activation of triphenyl-sulfonium hexafluoroantimonate to induce hydrogen abstraction from γ-butyrolactone, in which the photoresist SU-8 is commonly dissolved. The as-formed fluoroantimonic acid is the required initiator for the SU-8 crosslinking, shown in Panel (B).

The last-described steps are the most central ones of the whole PL process. In that context, some comments are adequate on the type of equipment needed for them. In general, one may use a simple self-built facility consisting of not much more than a kind of picture frame with a quartz-glass plate, into which the resin-coated silicon wafer and the photomask are sandwiched on top of each other, such that this arrangement can then be placed underneath a strong UV lamp that provides homogeneous light of sufficient strength (approx. 10 to 40 mW cm^{-2}) over an area of about 100 cm^2. Much better than that, though, are professional mask aligners that come along with perfectly pre-tuned settings of substrate plus photomask placing and positioning underneath a high-end UV source. These mask aligners even allow multiple such steps to be conducted with precise photomask positioning in each, such that even multiple-height master devices can be built with micrometer-scale precision. These tools, however, are costly, ranging from approx. 50,000 to 300,000 Euros.

Looking at the lengthy, multistep microchannel master device fabrication by PL, it seems that the manufacturing process requires careful planning of the working day, and that it renders the overall microfluidic device fabrication rather inflexible. To account for this concern, one may split-up the PL process into single steps if the lab day is too packed. Indeed, PL can be separated into individual working packages, as long as certain steps are carried out in one go to avoid detriment of the resolution of the microfluidic master device. For instance, the process of PL can be paused after the soft bake of the resin as long as the resin-coated substrates are stored in darkness afterward. Yet, one should not take a break after the UV-exposure of the resin, as diffusion of the so-formed Lewis acid into nonexposed areas will significantly decrease the patterning resolution.

After the PL process, the final master device should be subject to structural characterization before further use. For this purpose, several methods of optical microscopy can be used, as compiled in Figure 3.12. As the microchannel architecture of a flow cell is tailor-made for each application, each device can significantly differ in height and complexity from another, complicating mutual comparison and identification of production flaws. It is thus suggested to include standardized design features in the layout of every microfluidic device at the same position for quality control. These structural features are not part of the actual microchannel network, but help the experimenter to quickly identify whether the master device – as received by the PL manufacturing – is over- or underexposed, whether the resolution of the photomask translates one-to-one into a desired structure regarding edge sharpness and minimal feature size, and whether the developer bath time has been sufficient to remove all unpolymerized resin. For that purpose, two types of quality control structures are suggested: one is made of lines that translate into high-aspect ratio, freestanding walls, and the other is the exact inverse of those, translating into deep grooves, as both shown in Figure 3.13.

3.1.5 PDMS-based microfluidics: Device replication by soft lithography

To mold a microflow cell from the previously fabricated master device (cf. Section 3.1.4), elastomers are commonly employed. As introduced in Section 3.1.1, **poly(dimethylsiloxane) (PDMS)** is the most widely applied material for that purpose due to its unique combination of favorable properties, including good optical transparency, preparatively tunable elastic modulus in the MPa range, cheap cost, and ease of preparation. PDMS elastomers can be made from a two-component system of PDMS oligomers and a crosslinker, whose chemical structures are shown in Figure 3.14A and B.[11] This

---

[11] The crosslinking of PDMS proceeds via addition of reactive Si–H bonds, which occasionally occur in the siloxane crosslinker shown in Figure 3.14B, to the vinyl-terminated PDMS base shown in Figure 3.14A, thereby yielding an ethylene-bridged silicone polymer network, as shown in Figure 3.14C.

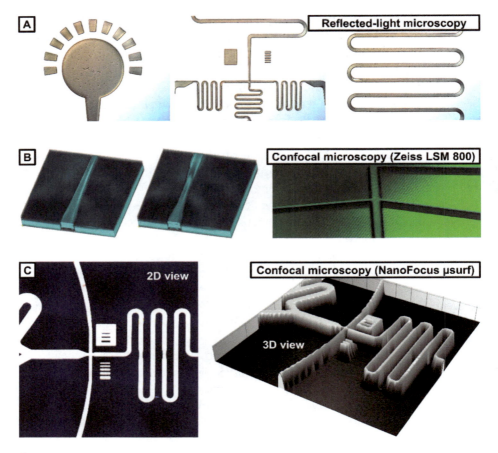

Figure 3.12: Common characterization techniques for analyzing microchannel master devices with regard to the presence of structural defects, dust particles, as well as in view of quantifying the microchannel height, width, and profile. (A) Reflected-light microscopy and (B) conventional confocal microscopy as well as (C) confocal imaging utilizing a system specifically designed for 3D surface metrology.

set of two components is commercially available under the name **Sylgard elastomer construction kit**. With such a kit, device replication can be conducted in a straightforward process named **soft lithography** (SL), as illustrated in Figure 3.15.

First, the PDMS oligomer base and crosslinker are mixed, typically in a ratio of 10:1; other ratios may be used to obtain harder (more crosslinker) or softer (less crosslinker) formulations if desired. This mixing is done in a total amount of typically 55 g (i.e., 5 g crosslinker plus 50 g PDMS base). Most optimally, the formulation is thoroughly mixed without creation of air bubbles in it though use of a Thinky orbital mixer, which is a small tabletop tool of extreme utility for this purpose, though costly (about 10,000 EUR). The mixed PDMS formulation is then

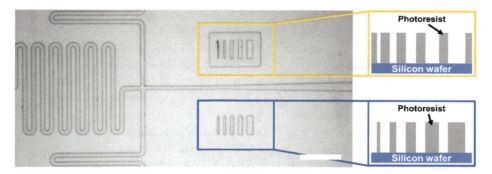

Figure 3.13: Quality control of the lateral resolution in photolithography by standardized design features. Two boxes with elongated, microchannel-mimicking wells (orange box) as well as freestanding photoresist walls of the same dimensions (blue box) are employed to check for correct baking and illumination times employed during the PL process, and to check for sufficient resolution of the photomask. Note: to avoid any interference with the flow inside the microfluidic device, these design features should not be part of the actual microchannel network. The scale bar denotes 500 µm.

Figure 3.14: Chemical structure of components conventionally used for preparing stamped microfluidic devices by PDMS-based microchannel replication. (A) Vinyl-terminated polydimethylsiloxane, (B) siloxane crosslinker (R = Me and occasionally H), and (C) crosslinked PDMS.

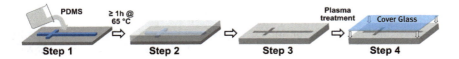

Figure 3.15: Replication of a microchannel master device via soft lithography (SL). In Step 1, PDMS oligomer and crosslinker are mixed, commonly at a ratio of 10:1, and then poured onto the microchannel relief obtained previously by PL. In Step 2, this PDMS formulation is crosslinked at approximately 65 °C for 1 h and then, in Step 3, peeled-off from the master structure, thereby obtaining a negative imprint of the channel structure in a rubbery slab of PDMS elastomer. Before further processing, fluid in- and outflow ports are punched into the replica (not shown here), and then, in Step 4, the PDMS replica is washed to remove dust and PDMS debris, dried by compressed air or nitrogen, and then sealed with a cover glass slide after air or oxygen plasma treatment (cf. Section 3.1.6).

poured onto the microfluidic channel master obtained from the previous PL step (which may be long ago, as each master can be used for long time in numerous SL replication steps), as shown in Step 1 of Figure 3.15. For this purpose, it is most handy to place the disk-shaped master (which has been patterned onto a disk-shape silicon wafer, as outlined in Section 3.1.4) into a Petri dish of matching size and then pouring about half of the 55-g PDMS formulation into it; the other half can be used for a second master in a separate Petri dish. After that step, it is crucial to remove air bubbles from the liquid PDMS; this is best achieved by placing the Petri dish (plus some others that are in the same process) into a plastic desiccator connected to a standard rotary vane pump that creates reduced pressure (not a strong vacuum, but just reduced pressure), thereby causing foaming of the residual air in the PDMS. Quick venting of the desiccator may then break this foam, thereby releasing the air from the PDMS. This procedure may be repeated several times. If after that, some last nasty air bubbles still remain in the PDMS formulation, these may be removed by blowing air pulses onto them using hand bellows. The Petri dish is then placed into an oven at 65 °C for 1 h, thereby hardening the PDMS by crosslinking it into a structure as shown in Figure 3.14C, illustrated as Step 2 in Figure 3.15. After cool-down to room temperature, a slab of PDMS that contains all channel structures wanted is cut out of the hardened elastomer and peeled off for further use, and the hole that this leaves is re-filled with new PDMS formulation and then baked again at 65 °C for h for storage of the master. To cut out the PDMS slab, it is best to use a scalpel, keeping a safe distance of a few millimeters to the microchannel features to maintain some surrounding PDMS material that will be needed for substrate bonding later. By peeling off the cut-out piece, a rubbery PDMS slab is obtained that features negative imprints of the channel structure, as illustrated in Step 3 of Figure 3.15. This slab is used in a further procedure of processing steps to eventually obtain a microflow cell, shown as Step 4 in Figure 3.15 and detailed in the following.

3.1.6 PDMS-based microfluidics: Assembling microfluidic devices from microchannel replicas and substrates

Having obtained a PDMS slab containing the desired microchannel network by soft lithography replication, the final microfluidic device lacks two further features. First, the microchannel structure needs a fundament, and second, it also needs access points to inject and exhaust fluids. To address the latter and to become able to inject and harvest fluids into and from the microfluidic device, we require in- and outflow ports, into which we can plug (flexible) tubing. For that, holes are punched into the PDMS slab at prescribed positions for the fluid in- and outlets (the design features for these have been discussed in Section 3.1.3) employing a biopsy punch needle. Although these needles have been originally designed for extracting cylindrical tissue samples, they are also suitable for punching holes into rubbery PDMS. In this process, it is crucial to precisely position and punch the in- and outflow ports exactly over the microchannel, for what we have discussed helpful channel design features in Section 3.1.3, and to use new and sharp biopsy needles.[12] Only then, the experimenter will be able to interface the microchannel network with an external fluid pumping system (cf. Section 3.2.2) without leakage. This interfacing is done by simply hand-plugging-in tubing into the rubbery PDMS, typically polyethylene laboratory tubing with an outer diameter slightly larger than that of the punch holes (whose size is given by the type of biopsy needle used to punch them). The compliance of the PDMS material will allow the plug-in of the tubing to be done quite easily, and it will also ensure that the tubing will stay in place afterward, even during fluid supply.

In the next step of device production, the PDMS replica with added fluid in- and outflow ports needs to be cleaned from any dirt and PDMS debris that has been created during the hole punching, with particular care to the circumstance that such debris often adheres to the replica's surface. This is best achieved by first sticking Scotch tape to the PDMS and then ripping it off, thereby removing larger particles and debris. After that, rinsing with high-quality (such as HPLC-grade) isopropyl alcohol is recommended. To remove that solvent in turn, the slab is subject to air or nitrogen blow. In that step, it is crucial to ensure that the blow-gas is oil free, which may be an issue in laboratories where oil finds itself in the gas pipelines. If such oil residues in pipeline-provided compressed gas cannot be ruled out completely, it is advised to use compressed synthetic air or nitrogen from a gas bottle supply instead.

To this point, the microchannel structure is still open to air. To seal it, we finally bond the whole replica to a substrate. This substrate should be chosen to support the squishy PDMS slab and, of course, be transparent to allow for simple characterization of the microfluidic device and later experiments in it, for example, by light microscopy (cf. Section 3.2.4). Thus, glass substrates of the size of standard microscope slides (approx. 75 mm × 25 mm) or thin ones as used for fluorescence microscopy at low working distances are most suited. We recommend to use as new as possible, recently unboxed glass slides, but even then, the surface of these slides can be contaminated with a thin greasy film that some manufacturers use to keep the glass slides separated from each other for easier handpicking. We recommend to not use such treated glass slides, but pure native ones instead, although it may be nasty to separate them from one another in an as-received packed box. A good way to quick-test if the intended glass slide is good or not is to pour some drops of water onto it. If the water wets the glass, then it is good, but if the water does not wet, then it is bad.

---

12 Biopsy needles are actually designed as single-use items to take samples from tissue. Nevertheless, they can be used to punch holes into soft PDMS multiples times. In this process, the needle will blunt over time (approx. after 50 to 100 punches), which can be observed by the occurrence of characteristic cracks within the PDMS that laterally extend from the punch hole, and that are responsible for leakage and loose tubing later. Then, the experimenter just needs to sharpen the tip of the biopsy needle by treating it with sandpaper or a deburrer to extend its lifetime.

On top of that, it is further recommendable to thoroughly clean the glass slide by rinsing with high-quality (such as HPLC-grade) isopropyl alcohol that is then removed by air or nitrogen blow, again taking care of doing that with oil-free gas!

Finally, to achieve a tight connection of the glass substrate and the channel-imprinted PDMS slab, plasma treatment of both parts aids to permanently bond them together. In the process of plasma treatment, both the glass substrate and the PDMS slab are exposed to an ionized gas (plasma) within an ignition chamber (cf. below for details). Depending on the gas source used, as the plasma is ignited, the flow cell building parts become surrounded by a cloud glowing in purple (argon), lavender (oxygen), or pink (nitrogen). During reaction of an air or oxygen plasma with the surfaces of the microchannel building pars, silanol groups are generated on both the glass and the PDMS. These may undergo condensation reactions with one another, forming covalent Si–O–Si linkages that irreversibly bond both flow cell parts together.[13] To put this into practice, both the cleaned glass slide and the PDMS slab (in a supine position) are placed into a plasma cleaning oven (in short: "plasma cleaner") and subjected to quick plasma treatment, as detailed in the following info boxes. After that, they are sandwiched on top of each other and hand-pressed together for some seconds. To finalize the silanol reaction, the resulting array of glass and PDMS is placed in an oven at 65 °C for about an hour. Note: such bonding can also be done onto another PDMS slab rather than onto a glass slide, thereby allowing multiple-layered PDMS devices to be made. Further note: a discussion of parameters for improving the bonding between PDMS and glass, and a list of tests for determining the quality of the plasma treatment follows in Section 3.5.

At this point, the microfluidic device is ready for use, and fluids can be pumped through the microchannels. However, additional processing may be necessary to chemically tune the PDMS surface properties. As will be detailed later, in Section 3.4.1, the ability to functionalize PDMS channels, ideally in a spatially resolved fashion, is essential when multiple emulsions shall be formed within planar flow cells. For simple W/O single-emulsion formation, however, no such additional microchannel surface modification is required, as PDMS is hydrophobic by default. Thus, an inner aqueous phase will de-wet from the PDMS microchannel walls and become encapsulated in a continuous oil phase, thereby ensuring emulsion formation with water as the dispersed phase and oil as the continuous phase. Yet, the contact angle of PDMS against an aqueous phase can be further advanced to even better minimize contact between the inner aqueous

---

[13] Plasma treatment is actually commonly applied as a method to clean surfaces, to delete chemical surface functionalization, or to even form a glassy layer on silicon surfaces (indicated by a colorful gleam at the surface) at high plasma powers. Therefore, it is essential to carefully balance the plasma power and exposure time to achieve sufficient formation of silanol groups but not yet glassy layers, because those would hinder the subsequent flow cell assembly.

Microfluidic device assembly by plasma-activated bonding:
General handling of plasma cleaners (Part 1)

As **plasma cleaners are usually not specifically designed for microfluidic device fabrication**, it is important to be familiar with key parameters of a machine, how to set them correctly, and to know their effect on plasma bonding efficiency / pressure resistance, bonding uniformity, and reproducibility. **Example**: **Low-pressure discharge plasma cleaner**, max. power: 100 W, chamber volume: approx. 2 L, gas source: oxygen bottle

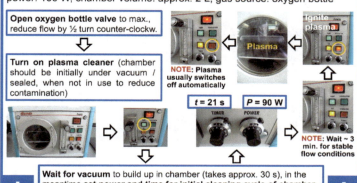

Microfluidic device assembly by plasma-activated bonding:
General handling of plasma cleaners (Part 2)

After performing initial **cleaning of the plasma chamber** (only necessary if started up for the first time of the day) **at elevated power and time compared to the later bonding processes**, the plasma chamber is ready to be loaded with PDMS replica and glass substrates.

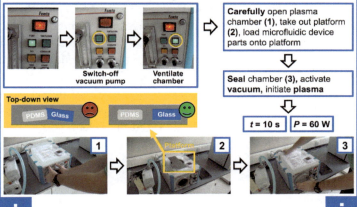

> **Positioning a PDMS-slab replica on a substrate after plasma activation**
>
> Positioning of a PDMS replica on a substrate (e.g., for preparing multi-layered flow cells) is tricky. If both the plasma-activated PDMS and the substrate (either glass or a second PDMS slab) are kept in contact only for a few seconds, they may not be separable anymore. And even if the experimenter would (rather brutally) rip off the PDMS and the glass again, their **surface properties would have changed during their time of contact**. Thus, it will be challenging to achieve sufficient bonding in a second attempt. To be able to move a PDMS replica on a surface for spatially precise alignment with the substrate *and* preserved plasma activation, **pipette approx. 10 μL water or ethanol onto the PDMS / glass surface right after plasma activation to form a liquid film**. On this film, the PDMS replica can be easily slid. As the solvent layer diffuses into the PDMS replica and evaporates, respectively, desired permanent bonding occurs.

phase and the microchannel wall behind the droplet-forming nozzle by a very simple means: the microfluidic device can be treated with **Aquapel** (Pittsburgh Glass Works LLC), a commercial glass treatment agent for fluorinating surfaces. During such a treatment, the silane-based, humiditiy-sensitive reactive solution adsorbs to the microchannel surface and diffuses into the rubbery-elastomeric PDMS material, thereby creating a permanently fluorinated PDMS-channel surface that is very hydrophobic and therefore very well suited for W/O emulsification in SF microfluidics.[14]

Having completed the whole procedure outlined on the past pages has led us from a virtual drawing of a desired channel structure, as shown in Figure 3.16A, over a PL-made master device, as shown in Figure 3.16B, to a SL-made PDMS replica, as shown in Figure 3.16C. The latter can be reproduced many times from the same master by just repeating the SL-part of the process, with practically no significant variation from one copy to another. The usual death of the master device that is used as a mold for PDMS replicas is accidental dropping and breaking in a moment of inattention in the lab, but usually not due to inherent ageing or degradation.

So far, our PL and SL procedure has led us to microflow cells with channel structures of a single, uniform height. More advanced microfluidic experiments, such as double-emulsion formation in channels with no spatially patterned wettability, require

---

14 If microfluidics users have access to a synthesis lab with Schlenk line and argon supply, they may consider preparing reaction mixtures for fluorinated surface coatings on their own (see Chapter 3.4.2).

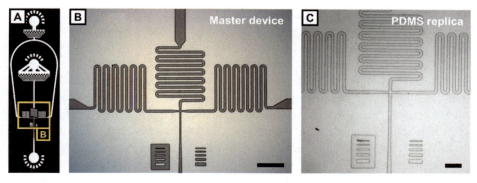

Figure 3.16: From a CAD drawing to a stamped microfluidic device. (A) Photomask developed in AutoCAD. White areas in the drawing appear transparent in the later photomask and allow for spatially controlled passage of UV-light onto a photosensitive surface, for example, a silicon wafer coated with a negative or positive resin. (B) Reflected light microscopy image of a photopatterned master device, here featuring a microchannel cross-section for emulsion formation. Dark features or microchannels are elevated above the silicon wafer substrate to allow for replication by soft lithography. Meander-like channels included in the flow cell design serve as resistors and prevent unintended backflow of the continuous and dispersed phases, respectively, during the flow cell start up in later experiments. Marks in the lower part of the image visualize the resolution limit of the photolithography process. The scale bar denotes 200 µm. (C) Replica of the flow cell's master device of (B) in a PDMS silicone elastomer bound to a glass substrate. The scale bar denotes 500 µm.

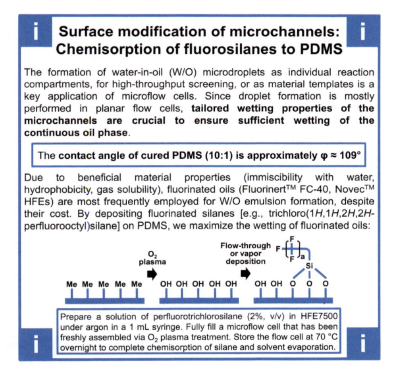

the use of nonuniform height channels, for example, channels that abruptly gain in height behind a cross-junction to prohibit fluid-to-wall contact of an inner, to-be-dispersed phase. Such devices can be fabricated in a slight variant of the process. The key step in this variant is to fabricate two master devices with multiple heights, one made such to mold an upper half of the later channel system and one to mold its lower half, and then to bond two complementary PDMS replicas made from these masters together such to obtain channels with multiple heights. Such master devices can be made by multistep photoresin-curing, hardening, and removal. In a first step, a first layer of the resin is coated and soft-baked onto a silicon wafer and exposed to UV light through a first photomask featuring the ground-level part of the channel structure, but this layer is then not yet developed, but instead, a second layer of photoresin is coated onto that and then subject to a secondary UV-illumination step such to harden the second-level part of the channel structure on top of the first. This may be repeated further times if desired, until eventually, all the unexposed parts of the resin are taken away by development. The result is a master device featuring a positive relief of a desired channel structure with parts of different stepwise height. In that sequential part-by-part photo-patterning procedure, it is crucial to ensure precise placement of the different photomasks used in the different patterning steps on the substrate. To achieve this, the photomasks must all be plotted with the absolute same scale, and they must be equipped with marker labels on the corners that can then be brought to precise overlap in each UV-illumination step using a mask aligner. Furthermore, it is beneficial to equip the top- and low-half masters that are made in the PL part of the process with complementary edge features such as pins and complementary holes to allow the two complementary PDMS replicas that are made in the SL part of the process to be sandwiched together in the right alignment in the step of plasma bonding.

The fabrication of microfluidic devices detailed to this point is based on the building materials PDMS and glass. In anticipation of Section 3.2.4, which will focus on optical inspection and characterization of microfluidic experiments, flow cells made of these materials are perfectly suited for any optical or spectroscopic characterization method involving visible light and IR radiation.[15] But how about characterization methods utilizing radiation with higher energy and lower wavelength, respectively? Here, we face the limitations of PDMS-and-glass microfluidics. To analyze flow patterns and reactions within microfluidic devices, thus, other building materials are needed that exhibit both low X-ray absorption and X-ray scattering. One of these materials is **Kapton**®, a polyimide made by DuPont. Kapton® is widely commercially available as a self-adhesive tape and can thus easily be employed to seal open microchannels made by combined PL/SL, as detailed in the following info box [15].

---

15 A prerequisite for IR-based analytics is that the PDMS is fabricated into thin layers to serve as IR-transparent substrate.

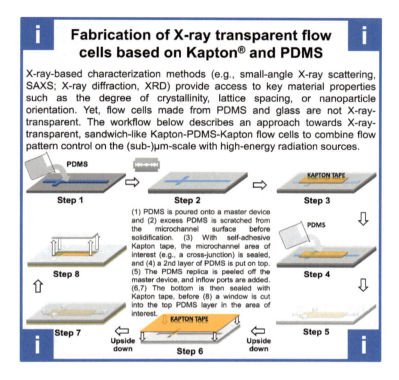

3.1.7 Glass-capillary microfluidics: Manual fabrication of flow cells

The inherent advantages of microfluidics based on stamped flow cells that are produced in a process of PL and SL, as outlined in the preceding chapters, is the ability to exactly reproduce the devices in large numbers in a standardized procedure. Also, in this variant of microfluidics, numbering up by parallelization of multiple channels, as detailed in Section 2.5, is straightforward to realize. Nevertheless, this method has drawbacks. The greatest flaw of PDMS-based microfluidics is its limitation to the use of fluids that are compatible with PDMS. This selection of solvents is narrow, as many organic media swell PDMS, including chloroform, toluene, and short alkanes, which are all useful in several processes (most prominently those involving polymers that only dissolve in these media). Actually, the only kinds of fluids reliably working in PDMS channels is long fluorinated oils, long alkanes (which comes along with elevated viscosity, though), and water-based liquids. If a desired application requires other fluids to be used, then PDMS microfluidics soon reaches its limits.[16]

---

[16] For applying other organic solvents such as chloroform, toluene or tetrahydrofuran (THF), microchannels in PDMS need to be improved regarding solvent resistance by glass-like, solvent-based coatings (cf. Chapter 3.4.1).

An alternative to the variant of stamped flow cell microfluidics based on silicone elastomers such as PDMS is glass-capillary microfluidics. In this variant, devices consist of **coaxial assemblies of microcapillaries**. One of the inherent advantages of these devices is that since they are made of glass, their wettability can be easily controlled by a surface reaction with an appropriate surface modifier. For example, a quick treatment with octadecyltrimethoxysilane will render the glass surface hydrophobic, whereas a treatment with 2-[methoxy(poly-ethyleneoxy)propyl]trimethoxysilane will make the surface hydrophilic. This is particularly useful inasmuch as the wettability treatment can be done with different construction pieces (i.e., with different glass capillaries) of a device *separately* before assembling them together, thereby easily allowing different parts of a complex channel system to be modified with different surface wettabilities, which is exquisitely useful in double- or higher-order emulsion formation. In contrast to the ease of this process in glass-capillary microfluidics, the same task is particularly puzzling in PDMS-based microfluidics, as detailed in Section 3.4.3. An additional benefit of using glass is that the devices are both chemically resistant and rigid. Lastly, and perhaps most importantly, these devices offer the distinct capability of creating truly **three-dimensional flows**, in which an inner fluid stream can be fully surrounded by an outer one and does not get in touch to the channel walls. This is highly beneficial, as with such a flow pattern, the need for channel-wall surface wettability adjustment is circumvented, which, by contrast, is of crucial necessity in narrow, planar stamped microflow cells, particularly in the variant of SF microfluidics. Such a flow pattern of a three-dimensionally enwrapped inner stream is also beneficial if fluids are to be used that are prone to induce surface fouling, which is often a detrimental issue in planar stamped flow cells. In view of all these latter advantages, there has actually been effort to mimic the beneficial three-dimensional geometry of glass-microcapillary devices also in PDMS stamped flow cells [16], but this relies on a sophisticated procedure involving multiple-height device masters (as just discussed above in the end of Section 3.1.6), whereas building such channel geometries from glass microcapillaries is easy and straightforward.

To build a glass-capillary microfluidic device, we begin with a circular glass capillary with an outer diameter of 1 mm, which is heated and pulled using a pipette puller to create a tapered geometry that culminates in a fine orifice. That capillary is then carefully inserted into a square glass capillary with an inner diameter of 1.05 mm. Coaxial alignment of the two capillaries is ensured by the match between the outer diameter of the circular capillary and the inner dimension of the square capillary. Fluid supply is enabled by placing syringe tips over the regions of the capillary arrangement where fluids are to be injected, which is the wide open end of the round capillary and the interstice between the inner round and the outer square capillary. Later on, in experiments, tubing can be plugged over these tips to interface syringes that are placed in syringe pumps for the fluid supply. As an outlet, we use a second round capillary that is plunged into the square capillary downstream, over whose open end another piece of tubing may be plugged. All the parts of this device are finally fixed on a glass substrate, and all slits and remaining interstices are sealed with five-minute epoxy glue. Figure 3.17A sketches the resulting co-flow glass-capillary device in a schematic fashion. Flowing one fluid inside the circular capillary while flowing a second fluid through the square capillary in the same direction results in a three-dimensional coaxial flow of the two fluids, as illustrated in Figure 1.26A–D in Section 1.5.2; this is known as **co-flow geometry**.

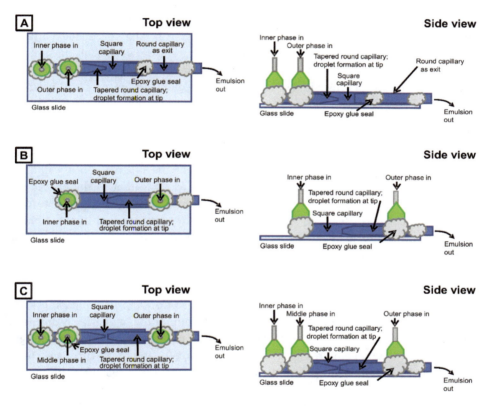

Figure 3.17: Three variants of hand-assembled glass-microcapillary devices. (A) Co-flow geometry for either CF microfluidics or for co-flow drop-formation in SF microfluidics. (B) Flow-focusing geometry, which is particularly useful for droplet formation in SF microfluidics, (C) Combination of both co-flow and flow-focusing, which is a powerful geometry to form double-emulsion droplets in a single step.

An alternate, differently arranged geometry, which is useful especially for drop formation in capillary devices, is the **flow-focusing geometry**. In contrast to co-flow capillary devices, two fluids are now introduced from the two ends of the same square capillary, from opposite directions. The inner fluid is hydrodynamically focused by the outer fluid through the narrow orifice of the tapered round capillary, as shown in Figure 1.26E in Section 1.5.2 as well as visible in Figure 2.8A in Section 2.2.3. Under dripping conditions, drop formation occurs as soon as the inner fluid enters the circular orifice, whereas under jetting conditions it occurs further downstream. An advantage of this method is that, in SF microfluidics, it allows us to make monodisperse drops with sizes smaller than that of the orifice. This feature is useful for making small droplets (~10–50 µm in diameter), especially those from a particulate suspension, where the particles may clog the orifice in the co-flow geometry. The use of a capillary with a larger orifice minimizes the probability of such tip clogging by the suspended particles or any entrapped debris. Figure 3.17B sketches such a flow-focusing glass-capillary device in a schematic fashion. In this variant of a device, fluid supply is again enabled by placing syringe tips over the regions of the capillary arrangement where fluids are to be injected, which is now the open end of the square capillary and the interstice between the square and the oppositely inserted round capillary. Yet again, in experiments, tubing can be plugged over these tips to interface syringes that are placed in syringe pumps for the fluid supply. As an outlet, this time, we may simply use the open end of the tapered round capillary, over which another piece of tubing may be plugged. Note that in this device geometry, wettability adjustment of the tapered round capillary is helpful to promote the emulsion formation, which should therefore be rendered hydrophobic if W/O emulsions are to be made, whereas it should be rendered hydrophilic if O/W emulsions are to be made, both achievable by quick dipping of the tapered end of the round capillary into a suitable silane reagent before assembling the microfluidic device.

If the two latter types of devices geometries – co-flow and flow-focusing – are combined, a device results that is suitable to forming double-emulsion droplets. At the tip of a first tapered round capillary that finds itself in a co-flow arrangement within an outer square capillary, a drop is formed, and at the tip of a second, larger, and frontally facing second round tapered capillary, that first drop is instantaneously encapsulated into a second, outer drop, thereby forming a double emulsion, as sketched in Figure 2.6 and as well as visible in Figure 2.8B, both in Section 2.2.3. Figure 3.17C illustrates such a double-emulsion glass-capillary device in a schematic fashion. Again, in this device geometry, wettability adjustment of the larger tapered round capillary is helpful to promote the emulsion formation, which should therefore be rendered hydrophobic if O/W/O double emulsions are to be made, whereas it should be rendered hydrophilic if W/O/W double emulsions are to be made, both achievable by quick dipping of the tapered end of the larger round capillary into a suitable silane reagent before assembling the microfluidic device.

The most severe disadvantage of glass-capillary microfluidics is its inability for numbering up. This is because the devices are fabricated by hand in no exactly reproducible fashion, which prohibits multiple identical copies of the same channel system to be made. In a non-numbered-up variant as a laboratory tool, though, glass capillary devices have great strength, robustness, and reusability, as they can be rinsed with cleaning fluids such as isopropyl alcohol after each use and then used over and over again.

To build glass-capillary devices, just some basic equipment is needed, including epoxy-glue syringes and disposable small plastic bowls for glue preparation, a collection of sandpaper of various roughness, a mini grinding machine, a simple optical microscope, and a small chemical workbench with standard cleaning and rinsing fluids such as water and isopropanol. There is just one large and costly item needed: a micropipette pulling machine (for example, a Flaming/Brown type micropipette puller P-1000 by Sutter Instruments, Novato, CA, U.S.A.). The following info boxes illustrate the manual procedure of glass-capillary microfluidic device preparation.

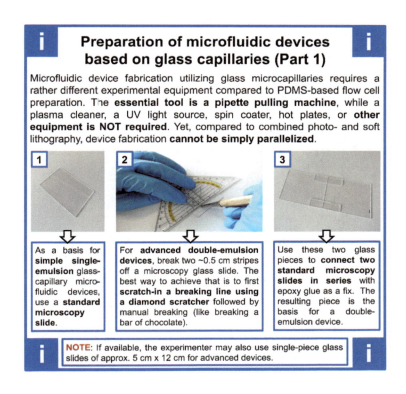

Preparation of microfluidic devices based on glass capillaries (Part 2)

4 Glass microcapillary with squared cross-section

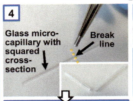

Break line

5 Single-emulsion device basis / Double-emulsion device basis

Place the resulting half-piece of the square capillary onto the device base / glass slide.
Single-emulsion device: standard microscopy slide
Double-emulsion device: self-made extended slide

Next, take a **commercial square-cross-section glass microcapillary**, with an **inner diameter of 1.05 mm**, and scratch a breaking line into its middle using the diamond scratcher again. Break the capillary along that line into two shorter pieces.

6 Capillary / Heating filament (top view)

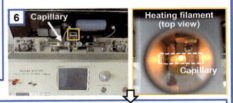

Now, take a commercial glass capillary with a **round cross-section and an outer diameter of 1 mm** and **place it into a pipette pulling machine** to **create two halves with tapered orifice each**. This is done by a heating filament followed by controlled capillary pulling, all performed in a programmable fashion by the pulling machine.

Preparation of microfluidic devices based on glass capillaries (Part 3)

7 Rub / Air-blow

Take one of the two resulting pulled capillary halves and extend its tapered orifice by **manual sandpaper rubbing followed by air blowing to remove glass dust**. A good technique to achieve clean glass edges is to twist the capillary while rubbing it very gently over the surface of sandpaper that is stretched in a frame.

8 Capillary after pulling / Capillary after sandpaper treatment / 200 μm

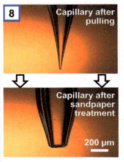

During the sandpaper treatment, occasionally **check the change of the orifice diameter** by optical microscopy. Usually, the virgin capillary that is obtained after pipette pulling has an orifice diameter of **just about 5 to 10 μm**, whereas after sandpaper treatment, the diameter will be **around 50...300 μm**, adjustable as desired simply by extending the sandpaper rubbing. For a **single-emulsion device**, you just need to have a large post-sandpaper capillary. For a **double-emulsion device**, by contrast, you need to have both, a narrow capillary as obtained right after pipette pulling and a broader one as obtained by sandpaper treatment. Ideally, the two orifices should have a **width-ratio of 1:5 to 1:10**.

Preparation of microfluidic devices based on glass capillaries (Part 4)

9 After rubbing the capillary orifice to its desired size, **dip it into a reagent for surface modification**. If a **water-in-oil emlusion** is to be made, this reagent is **octadecyltrimethoxysilane**; if an **oil-in-water emulsion** is to be made, this reagent is **2-[methoxy(poly-ethyleneoxy)propyl]trimethoxysilane**. This dipping will make the **silanes react with the glass surface** and cover it either with **hydrophobic octadecyl** or with **hydrophilic poly(ethylene oxide) chains**. This will ensure that only the **respective outer phase in the microfluidic experiment preferably wets the capillary walls, but not the inner phase**, leading to desired droplet formation. After dipping, remove the excess reagent by purging the capillary with either chloroform (in case of hydrophobic surface treatment) or with water (in case of hydrophilic surface treatment), followed by further flushing with isopropanol, and then drying by air flow.

Preparation of microfluidic devices based on glass capillaries (Part 5)

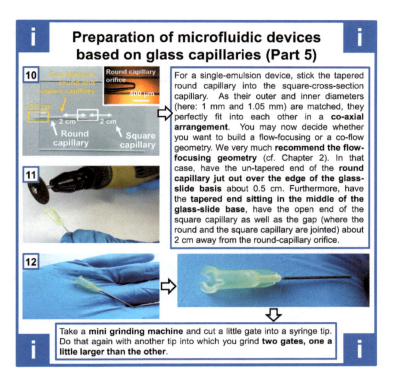

10 For a single-emulsion device, stick the tapered round capillary into the square-cross-section capillary. As their outer and inner diameters (here: 1 mm and 1.05 mm) are matched, they perfectly fit into each other in a **co-axial arrangement**. You may now decide whether you want to build a flow-focusing or a co-flow geometry. We very much **recommend the flow-focusing geometry** (cf. Chapter 2). In that case, have the un-tapered end of the **round capillary jut out over the edge of the glass-slide basis** about 0.5 cm. Furthermore, have the **tapered end sitting in the middle of the glass-slide base**, have the open end of the square capillary as well as the gap (where the round and the square capillary are jointed) about 2 cm away from the round-capillary orifice.

11

12 Take a **mini grinding machine** and cut a little gate into a syringe tip. Do that again with another tip into which you grind **two gates, one a little larger than the other**.

Preparation of microfluidic devices based on glass capillaries (Part 6)

13 Place the **grinded syringe tips on top of the parts of the device where fluids must be injected**. These are the open end of the square capillary and the little gap between the square and the round capillary.

14 Advance the device to a table edge and look from below to **ensure good placement of the syringe tips**. Ideally, they should centrally cover the fluid inlets.

15 **Fix and seal everything with epoxy glue**. Ideally, apply a freshly-mixed, low-viscous epoxy glue from the sides first to hold the syringe tips in place, and then, after a few minutes, when the epoxy glue gets more viscous and approaches solidification, also apply it over the grinded gates in the syringe tips such to ensure that it seals them, but **does not flow into them too deeply, thus prohibiting clogging of the device**. Most ideally, the **round capillary end should stand out of the glass-slide base for about 0.5 cm**. Fix that arrangement with epoxy glue as well.

Preparation of microfluidic devices based on glass capillaries (Part 7)

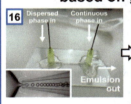

16 With that, your **single-emulsion device is done**. It can now be fed with a continuous external phase through the inlet on top of the capillary gaps, whereas a dispersed inner phase is injected into the inlet on top of the square-capillary open end. The **fluids will then frontally hit each other at the capillary tip**, where the inner fluid is dispersed by the outer fluid via flow focusing. The resulting emulsion will exit through the round capillary, over whose end a piece of tubing may be put to collect the emulsion in a collection vial.

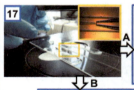

17 In case of a **double emulsion device**, stick both a narrow-tapered round capillary and a broad-tapered, surface-modified round capillary into a square-cross-section capillary, one from each end, respectively. The perfect match of the inner and outer diameters of the squared and round capillaries (1.05 mm and 1 mm) will create a perfect co-axial alignment.

Place the whole arrangement onto a microscope stage and **carefully align the capillaries such that they frontally face each other in a distance of about 0.5x to 1.5x the narrow capillary's orifice**. Furthermore, arrange the three capillaries such that the open un-tapered end of the broad one juts over the edge of the glass-slide basis about 0.5 cm. Use epoxy glue to fix that arrangement. Do that fixing on the microscope stage, as little post-adjustments of the alignment will be necessary during epoxy-glue application and drying.

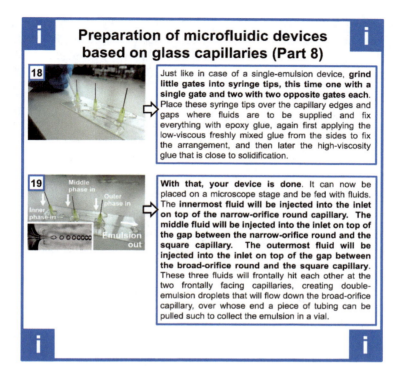

3.2 Running microfluidic experiments

3.2.1 Sample preparation for microfluidic experiments

Microfluidic devices can be fed with a myriad of fluids, including gases, suspensions of cells, vesicles, or nanoparticles, and liquids ranging from native (organic) solvents over aqueous solutions to oils. In view of the latter, when preparing liquid samples for microfluidic experiments, it is **crucial to aim for complete dissolution of all compounds**. Only then, microchannel clogging due to agglomerates and varying viscosity due to inhomogeneous distribution of solutes and resulting concentration gradients complicating the flow behavior are prevented. As we have previously discovered in Chapter 1, dissolving compounds – particularly polymers – can be challenging. While dissolving low-molar-mass compounds is a simple task, especially if a compound exhibits a high solubility in the respective solvent, an experimenter may have to wait minutes to hours or even longer for dissolving high-molar-mass species. On this account, experimenters often try to speed up the process of dissolution, for example, by shaking and vortexing.[17] While this kind of agitation usually preserves the molecular structure or correct folding state of compounds such as proteins, other techniques such as ultrasound should be treated with caution, as further detailed in the following sticky note.

Dissolving macromolecules: Polymer degradation by mechanical force

While **dissolving low-molar-mass substances is usually fast** or can be easily sped up by simply shaking the liquid sample, **dissolving high-molar-mass substances (e.g., polymers) can be challenging and time consuming**. To speed things up, experimenters often employ mechanical forces that go far beyond simple shaking to achieve complete dissolution of a macromolecular sample, e.g., employing ultra-sound (US). Yet, while treating hard-to-dissolve (polymer) samples in an ultrasonic bath may yield an obviously homogeneously mixed solution ready to be applied in microfluidics, one should **scrutinize if ultra-sound leaves a high-molar-mass species unchanged**. In fact, ultra-sound promotes formation of tiny bubbles with high temperatures and pressures. As these bubbles break, extreme local **forces may degrade high-molar-mass substances into smaller pieces**. These may dissolve faster, yet intrinsic properties of the macromolecule have drastically changed by then. Thus: Polymer solutions should not be treated in an ultrasonic bath to quicken-up the process of dissolution.

Once the liquid sample preparation is completed and fluids are ready for injection into the flow cell, experimenters may face another challenge, which is, yet, a key strength of microfluidics. Flow cells allow for handling low sample volumes in the microliter range, which can be processed into fluid flow patterns with individual compartments down to a few femtoliters, for example, microdroplets. Considering the **dead volume** of typical connections (e.g., polymer tubing) between a fluid reservoir (e.g., a syringe) and a microfluidic device as well as the dead volume of a flow cell itself, a few microliters of sample may literally vanish inside the experimental setup without a chance of forming the desired flow pattern. Using a syringe with needle and tubing representing one of the most common fluid reservoirs in microfluidics (cf. Section 3.2.2), a few microliters of liquid sample can still be quantitatively injected into a microfluidic device by directly absorbing the fluid sample into the tubing instead of prefilling the much larger syringe body. By pulling the syringe piston while the tip of the tubing that is mounted

---

17 Experimenters should be particularly careful when trying to dissolve or homogenize biological samples as well as interfacially active substances such as surfactants or lipids. In the former case, mechanical stress may harm delicate biological assemblies like mitochondria, vesicles, or whole lysates, while in the latter case, shaking may introduce air bubbles that prevent stable flow inside the microfluidic device.

onto the syringe is submerged in the fluid sample, the fluid reservoir fills from the tip of the tubing toward the syringe body, and dead volume is minimized. However, as during the microfluidic experiment, the fluid reservoir is emptied into the opposite direction, the experimenter typically requires a few microliters to stabilize the flow inside the microfluidic device; thus, the fluid reservoir will be empty by the time the microfluidic device is ready for the actual experiment. Moreover, the user cannot receive liquids from the syringe on demand as the syringe body is filled with air, and only the tubing (and maybe the needle) contains fluid. Thus, as the fluid reservoir is actuated, the air is first compressed before the fluid starts moving inside the tubing. To overcome these two major limitations of air-filled syringes operated with low fluid volumes, the air inside the syringe body should be replaced by an incompressible, nonmiscible liquid. For that purpose – in case of a low-volume aqueous solution – both the syringe body and the connected tubing are filled with a (fluorinated), high-density oil first. Then, the aqueous phase is carefully sucked into the tubing. To avoid emulsion formation within the syringe during this process, it is crucial to hold the syringe with the needle up and the piston down. That way, the oil phase will remain at the bottom of the syringe with the aqueous phase on top. As the aqueous phase is inside the tubing, another plug of plain water separated by a short oil plug is sucked into the tubing. This aqueous phase will help to set up optimal flow rates without wasting the low-volume aqueous phase of interest that reaches the flow cell thereafter.

Despite the afore-mentioned steps that have to be undertaken for the preparation and injection of fluids into microfluidic devices, molecular solutions are rather easy to handle. Colloidal suspensions, by contrast, are a bit trickier to handle, as they pose the risk of clogging microchannels; on top of that, it is challenging to keep colloids such as cells, nano-, and microparticles homogeneously mixed, because they are prone to settle inside a syringe or other fluid reservoir. To prevent colloids from sedimenting on the timescale of a microfluidic experiment in academia (approx. 1 h to 1 day), we must modify conventional fluid reservoirs such as flasks, syringes, (Falcon) tubes, or pressure vials. While these are static and not able to hinder colloids from sedimenting, experimenters may construct a motor-driven magnetic stirrer that carefully agitates the colloidal suspension in a continuous fashion, thereby keeping the colloids floating. For a first trail, the experimenter may simply employ a small stirring bar that is moved by hand via an external stir bar retriever[18] inside the syringe body. A more sophisticated approach is to induce the motion of the stirring bar not by hand but by an external agitator, such as a small motor that rotates a small hand magnet in close proximity to the piece of the microfluidic setup that contains the stirrer. Such an agitator be assembled manually with pieces from a hobby-electronics supplier. Note: when utilizing a magnetic stir bar, it is important to carefully move it through the fluid reservoir to avoid change of the fluid flow.

---

18 A PTFE stick carrying a magnet at its tip to remove stir bars from flasks in chemical setups.

3.2.2 Moving fluids through microfluidic devices

Based on the governing equations of fluid flow, as discussed in Section 1.1.3, we can now focus on how to actually **generate and control fluid flow in microfluidic devices**. Transport of fluids within microfluidic devices can be achieved by **pressure-driven flow** induced by pumps, **entropy-driven flow**, which is basically diffusion, **gradient-induced flow** due to inhomogeneous temperature or concentration distribution within a microchannel system, and fluid flow induced by electric fields, such as **electro-osmosis**. Among these different technologies, syringe pumps are the most established flow control systems in microfluidics. The basic principle of this type of fluid flow control is illustrated in Figure 3.18A. As complex fluid flow patterns such as microdroplets often contain just a few picoliters, experimenters may choose high-precision syringe pumps, which can reproducibly pump as little as a few femtoliters per second. With these pumps, pulsatile flow that has been traditionally associated with syringe pumps has become rather negligible. For choosing and monitoring flow rates, the simplest syringe pumps (primarily intended for clinical use) contain a digital display, while more advanced pumps have integrated color touch screens that may simplify setting up and controlling flow experiments for new microfluidics users (e.g., Pump 11 Pico Plus Elite, Harvard Apparatus). If more complex pumping modes are required – for example, to allow for switching between draining and restocking with defined (such as sinusoidal) patterns – it may be required to employ an additional external syringe pump controller and computer, respectively (e.g., neMESYS pumps by Cetoni GmbH). While the main advantage of syringe pumps is that they are easy to use, their main drawback is a slow response time when setting a new flow rate value. Here, reaching an amended flow rate can take seconds to hours in the worst case. Moreover, if rather simple drive motors are built in, flow oscillations may occur caused by the motor steps. In addition, while positive-displacement syringe pumps are widely employed for feeding single microfluidic devices, they are rather inconvenient for parallelized operation of flow cells owing to the maximum limit of syringe sizes that can be plugged into such pumps. With a typical volume limit of approx. 50 to 60 mL (e.g., neMESYS 290N, Cetoni), microfluidic devices designed for scale-up will run out of fluids within minutes, such that a continuous long-term operation of them is impossible.

In view of these potential shortcomings of syringe pumps, an experimenter may consider pressure-driven fluid flow in a microfluidic setup instead. Pressure-controlled microfluidic setups allow for handling up to several liters of fluids in a single reservoir, thereby laying the foundation for scale-up of microfluidics. In contrast to syringe pumps, which provide direct flow control by movement of the syringe piston, a setup for pressure-controlled fluid flow consists of a pressure source to produce a gas input pressure within a connected liquid tank that is hermetically sealed, as sketched in Figure 3.18B. The pressure exerted onto the surface of the liquid within the reservoir induces a liquid flow from the tank to the connected microfluidic device. A flow controller monitors the fluid flow output versus pressure input and is able to regulate the

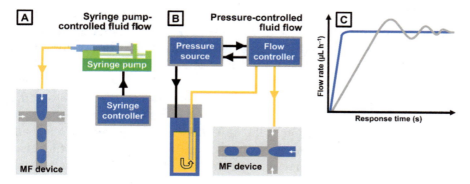

Figure 3.18: Comparison of fluid flow control in microfluidics. (A) Schematic of an experimental setup involving a syringe pump for flowing liquids into a microfluidic device. Depending on the choice of the syringe pump, flow-rate control may require an additional syringe pump controller and computer, respectively. (B) Schematic of an experimental setup for microfluidic flow regulation via (air) pressure. A controller pressurizes the surface of a fluid of interest contained inside a reservoir (e.g., Eppendorf tubes, Falcon tubes, glass vials), which is then injected into the microfluidic device by the inflow of gas. An additional flow controller allows for setting a desired flow rate, which is automatically translated into a corresponding pressure. (C) Stability and response time of pressurized flow systems (blue curve) compared to syringe pump-based flow control (gray curve).

input pressure to achieve constant flow until the fluid reservoir is empty. With this setup, pulseless and stable flow can be realized in a reproducible fashion within subsecond response time, as illustrated qualitatively in Figure 3.18C. Despite the multiple components required for setting up pressure-controlled fluid flow, the costs are in range of those for acquiring conventional syringe pumps. Note: with the increasing efforts toward systems integration in microfluidics, pressure-controlled fluid flow may also serve as a basis to develop functional integrated micropumps.

While pressure-driven flow allows for handling fluid volumes that exceed those of syringes mounted onto pumps, these volumes may still not be sufficient for large-scale operation of microfluidic technology in industry, which typically demands cubic meters, that is, tons of liquids. On this account, liquid pumps such as peristaltic pumps – similar to those used for aquariums – can be connected to indefinitely large fluid volumes. Unfortunately, they are usually pulsating during operation owing to the rotating pump head. In SF microfluidics, this flow behavior may lead to inconsistent droplet formation and large size variations, while in CF microfluidics, it may cause an unstable liquid–liquid interface and thus irreproducible reaction conditions.[19]

---

[19] Depending on the material basis of the microfluidic device, pressure fluctuations may be more or less distinct. For instance, flow cells made from glass, steel, COC, parylene, and other stiff materials do not damp flow fluctuations as efficient as soft, elastic building materials do. Along these lines, PDMS can provide sufficient compliance to damp remarked fluid flow pulsation originating from a pump, such that in these devices, in the case of SF, broadening of the produced droplet size distribution is insignificant.

All above-discussed methods for fluid flow control in microfluidic channels are based on pressure-driven flow, as discussed in Section 1.1. An experimentally more advanced technique for pumping fluids is that of **electro-osmotic flow** [17]. If the walls of a microchannel have an electric charge, as most surfaces do, an electric double layer of counter ions forms at the walls. When an electric field is applied across the microchannel, the ions in the double layer move toward the electrode of opposite polarity. This creates motion of the fluid near the walls and transfers into convective motion of the bulk fluid via viscous forces. A key advantage to electro-osmotic flow is that it is straightforward to couple other electronic applications on-chip. However, electro-osmotic flow often requires high voltages, making it a difficult technology to miniaturize without off-chip power supplies. Another significant disadvantage of electro-osmotic flow is the variability in surface properties. Proteins, for example, can adsorb to the walls, substantially changing the surface charge characteristics and, thereby, changing the fluid velocity. This can result in unpredictable time dependencies of the fluid flow, which renders electro-osmotic flow challenging to control in long-term experiments.

A fundamental difference between pressure-driven flow and electro-osmotically driven flow is the shape of the flow profile within a microchannel, as visualized in Figure 3.19. In pressure-driven laminar flow, the so-called no-slip boundary condition applies, stating that the fluid velocity at the walls must be zero (cf. Chapter 1). This yields a **parabolic velocity profile** within the channel, as shown in Figure 3.19A. Such a flow velocity profile has a significant effect on the distribution of molecules transported within a microchannel, leading to heterogeneous conditions within the fluid flow. This, in turn, diminishes a key selling point of performing reactions in microfluidic devices, which is normally given by the efficient mixing and homogeneous reaction volumes in microfluidics. Compared to that, in electro-osmotic flow, if the channel is open at the electrodes, and the velocity profile is uniform across the entire width of the channel, as shown in Figure 3.19B. However, sample dispersion in the form of flow banding that slightly broadens the distribution is still a concern for electro-osmotic pumping, as also seen in Figure 3.19B, yet leading to significantly lower dispersity in microfluidic product properties compared to a conventional parabolic velocity profile.

One major driving force in microfluidics research is devoted toward miniaturization and integration of valves, electrodes, mixers, heaters, switches, sensors, or actuators in microchannel networks. Yet, most microscopic devices require a cumbersome to-world connection to a pumping system to provide liquid flow to its microchannels. This limitation may hamper miniaturization and portability, as required for mobile analysis applications by end consumers, for example, for blood testing, food allergen detection, or immunoassays. Instead of using rather bulky external fluid pumps, an ideal, self-contained flow cell is able to provide fluid flow by

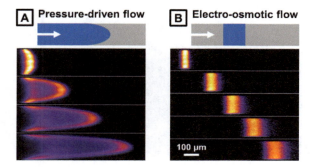

Figure 3.19: Comparison of flow fronts in microchannels utilizing pressure-driven (A) flow and electro-osmotically driven flow (B). The flow front is fluorescently labeled to visualize the sample dispersion. Ref. [18] presents a detailed study on these two kinds of flow profiles based on fluorescence labeling

Flow rate conversion "from microfluidics to SI units"

The SI unit for a flow rate is cubic meters per second $m^3\ s^{-1}$. However, microfluidic pumps often allow users to arbitrarily choose volume and time. Thus, the following conversion factors may help users to **swiftly compare flow rates** among experiments from different researchers and publications.

Non-SI unit	SI unit ($m^3\ s^{-1}$)
1 pL s^{-1}	1×10^{-15}
1 pL min^{-1}	1.67×10^{-17}
1 pL h^{-1}	2.78×10^{-19}
1 nL s^{-1}	1×10^{-12}
1 nL min^{-1}	1.67×10^{-14}
1 nL h^{-1}	2.78×10^{-16}
1 µL s^{-1}	1×10^{-9}
1 µL min^{-1}	1.67×10^{-11}
1 µL h^{-1}	2.78×10^{-13}
1 mL s^{-1}	1×10^{-6}
1 mL min^{-1}	1.67×10^{-8}
1 mL h^{-1}	2.78×10^{-10}
1 L h^{-1}	2.78×10^{-7}

itself. To reach this degree of system integration, **capillary forces** can be utilized to achieve autonomous, passive liquid transport [19, 20]. In this approach, fluid flow is governed by a capillary pressure exerted on a liquid that is simply loaded into a pad, open to air. The flow cell design contains all further information on velocity and distribution of the loaded liquid. Figure 3.20 presents an overview of different

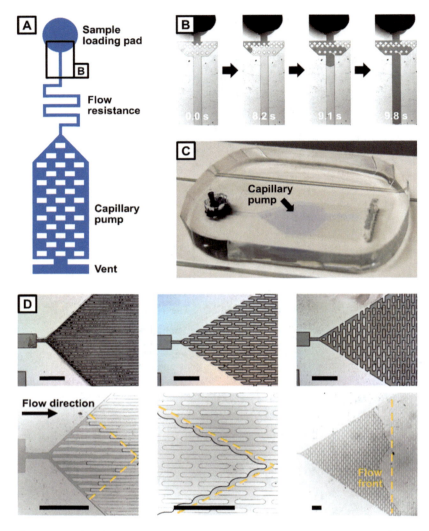

Figure 3.20: Fluid flow control in a microfluidic device utilizing capillary forces. (A) Schematic of a microflow cell for passively pumping a liquid through a microchannel network. The device consists of a loading pad from which liquid is dragged into the flow cell by capillary forces, a straight microchannel in which (bio-)sensing and analysis can be performed, a meander-like channel that serves as a flow resistance, and a large chamber with a vent acting as capillary pump.
(B) High-speed imaging sequence of the inflow of liquid into a microflow cell driven by a capillary pump. (C) Microfluidic device from (A) composed of a PDMS-based microchannel replica covalently bound to a glass substrate. (D) Filling behavior of microfluidic capillary pumps depending on the design and orientation of the geometric features within the pump. Top row: straight lines (left), high-aspect ratio posts along (middle) and perpendicular to the flow direction (right) prepared by photolithography on a silicon wafer, and visualized by reflected light microscopy; bottom row: corresponding PDMS-based flow cells filling up with an aqueous solution of Brilliant Black BN for enhanced contrast. The scale bars denote 500 µm.

elements and variants of these autonomous units. A key functional element in such a flow cell is a so-called **capillary pump**, which is basically a wide microchannel with large volume capacity, as shown in Figure 3.20A. These pumps spontaneously fill up with a speed in the range of nanoliters per second, as seen in Figure 3.20B, whereby no external power source or additional fluid flow controller is required. To control the loading of the capillary pump with a liquid sample from the loading pad, a meander-like microchannel acting similar to a resistor in electronic circuits (cf. Section 3.2) is slotted ahead of the capillary pump to modulate its filling process depending on the number of meander turns and thus the meander length. To efficiently fill up a capillary pump and make use of its maximum capacity, it is vital to optimize both the microchannel surface wettability and the microchannel aspect ratio. Only then, a continuous flow of liquid from the loading pad will form and reach the capillary pump to induce further fluid flow. In the former case, if an aqueous liquid sample is loaded into the device, the microchannel connecting the loading pad and the capillary pump needs to be highly hydrophilic; here the reader is referred to Section 3.4.1, where physicochemical microchannel surface treatments will be discussed. In the latter case, reducing the microchannel aspect h/w ratio from 1:1 to 1:10 will further improve the filling behavior of the microfluidic capillary pump device.

So far, moving fluids through microfluidic devices can be either achieved by actively pumping (using syringe pumps or pressure flow) or dragging fluids (using capillary pumps). While capillary pumps provide the means to design highly integrated, portable microfluidic devices suited for in-field analysis and characterization without any need for additional, occasionally expensive, equipment (cf. above), their design and operation requires profound knowledge of fluid-flow physics in microfluidic devices and how to tailor microchannel surface properties to achieve a set of desired flow rates. Capillary pumping is thus rather unsuited for users who are new to the field of microfluidics. Alternatively, experimenters may consider a method that is also based on dragging fluids through microfluidic devices, as previously introduced in Section 2.5, but that does not require integrated pumps. The technique could be literally named "microfluidics that sucks,"[20] as it utilizes an empty syringe, whose piston is pulled out and clapped to fix its position while being connected to a microfluidic device to exert depression on the microchannel network and the fluids therein [21].

In summary, microfluidic researchers mainly use pressure controllers when they require high flow responsiveness, flow stability, and precision, as well as when they work with dead-end channels or require large sample volumes. For users new to the field of microfluidics, positive-displacement syringe pumps are the pumping system of choice, being simple to set up and inexpensive. When choosing any of the discussed pumping systems – just like for any other equipment – it is vital to consider

---

[20] Similar to integrated capillary pumps, this technique is easy to set up and equally suitable to come up with portable, inexpensive microfluidic devices combined with fluid pumping.

the microfluidic setup's location, which does not have to be in a chemistry or physics lab at room temperature. Particularly, when handling sensitive biological samples or if a reaction is already triggered by the energy input provided by room temperature, the microfluidic device together with pumps, (high-speed) camera, and microscope may be moved to fridge-like temperature (approx. 4 °C), as found in a cold room, for instance.[21] For experiments at this temperature, it is important to ensure that any equipment used to assemble a microfluidic setup is certified to be operated in coldness. As standard computers may not be sufficiently cased or sealed to withstand both low temperature and high humidity as commonly found in cold rooms, experimenters should choose autarkic pumps and cameras (e.g., Phantom Miro eX4, Vision Research) that can be operated without external computers.

3.2.3 To-the-world connection of microfluidic devices

When setting up a microfluidic experiment, leakage and thus material loss needs to be precluded, as fluid reserves may be limited to a few microliters, especially in cases such as DNA replication or protein purification. In addition, to generate sufficient material output or to integrate a signal over a large amount of data, a microfluidic experiment may need to run for hours or even days without constant inspection, such that leakage must be safely excluded. For instance, to generate 1 mL of a suspension of polymeric hydrogel particles with 10 µm diameter from thermally polymerizable aqueous droplets, a flow-focusing device with 10 µm feature size and flow rates of 80 µL h^{-1} for the inner aqueous phase and 240 µL h^{-1} for the continuous oil phase will need to run 12.5 h nonstop. Choosing the right combination of tube sizes, diameters of the fluid in- or outlet ports, that is, punch holes (in case of stamped flow cells), and syringe needles is thus crucial.

To connect a microfluidic device to the "outer world," the outer diameter of the tube needs to match the diameter of the fluid in- or outlet ports and the punch holes (in case of stamped flow cells), respectively, to avoid leakage on the flow cell's side. If the device consists of a rubbery and compliant material such as PDMS, a best fit is

---

21 One may argue that fluid reservoirs that contain temperature-sensitive samples can be individually cooled by putting an ice-pack onto the syringe, or – in a more sophisticated approach – utilize flown-through cooling jackets or Peltier elements without jeopardizing functionality of microscope, pumps, and (high-speed) camera. Yet, syringe bodies that are actively cooled in direct contact with a cooling element may shrink at a different rate compared to the inner syringe piston material. Thus, the syringe becomes leaky during operation. Moreover, cool surfaces usually collect condensate that may drip into sensitive areas of the syringe pump. As simple light microscopes equipped with LEDs as light source hardly warm up during operation, high-speed cameras, as required for SF microfluidics, for instance, often fulfill military standards concerning their handling in a harsh environment, and syringe pumps are rather simple machinery that can withstand humidity and cold, a simple microfluidic setup may be built up and operated in a cold room, if available.

one where the tubing outer diameter is even a little larger than the fluid in- or output port's diameter, because then, the compliance of the surrounding PDMS will hold the slightly-too-large tubing tightly in place after its plug-in. On the other side of the fluid supply, the tubing's inner diameter should be slightly smaller than the feeding syringe needle diameter to ensure sufficient sealing and to avoid leakage there as well (especially at elevated pressure). In a typical scenario, a tube may have an outer diameter of 1.09 mm and an inner diameter of 0.38 mm; accordingly, syringe needles with an outer diameter of 0.45 mm are suitable, whereas the fluid in- and output holes in the rubbery device material should have a size of 1.0 mm.

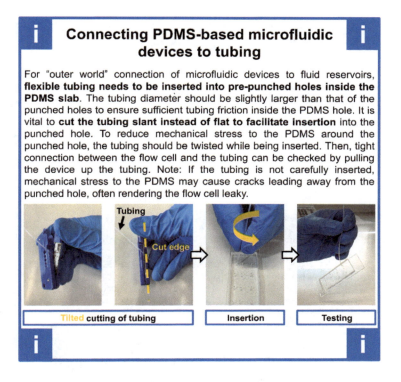

In addition to optimally defining the tubing specifications, one should keep the length of the tubing as short as possible to minimize dead volume and to reduce inertia – and thus response time – of the microfluidic experiment. For that, if conventional syringe pumps are used, it is advised to move the pumping system as close as possible to the site of observation (e.g., a bright-field microscope, as detailed further in Section 3.2.4); this can be further optimized when the syringe pumps are placed on stages of the same height as the microscope table.

When searching for a desired needle size, it turns out that the outer diameter of needles is mostly not given in metric units on syringe packages, but instead, as a so-called **gauge value** along with a color code. In this scheme, for example, a

Connecting needles to tubing (Part 1)

Microfluidic device operation is considered safe due to small volumes, controlled mixing, and efficient heat transfer, making fast or exothermic reactions easy to control. However, when setting up microfluidic experiments, users have to be precautious in one step: **connecting soft polymer tubing to a sharp, stainless-steel needle**. Since tight connection is required to prevent leakage, the outer diameter of needles is usually chosen to be slightly larger than the tubing's inner diameter, hindering smooth push-on of the tubing onto the needle tip. Yet, **two measures may help users** to connect their flow cells to syringes without the tubing or, even worse, their finger tips getting poked.

1. **Moving tubing onto a needle from below the needle tip:**

By putting tubing onto a needle in an angle corresponding to the cut of the needle tip (dashed line), penetration of the tubing wall is prevented.

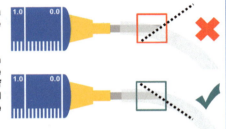

Connecting needles to tubing (Part 2)

2. **Keep both hands holding needle and tubing in contact**

When connecting needle and tubing, **one hand holds the tubing, the other one holds the needle**. Users usually poke themselves by sliding off when trying to push tubing onto a needle tip against friction forces between the needle surface and the tubing wall. Probability and severity of accidental skin penetration greatly depends on the force from accelerating the needle towards the user's skin upon impact.

If the hands holding tubing and needle are **centimeters away from each other**, the risk is high for the hand holding the needle to pick up sufficient speed to pierce the user's skin.

If both **hands remain in contact**, e.g., by the middle finger's tip, a user can stabilize the move and prevent accelerating one's hands towards each other, and thus prevent penetration of the skin by a potentially infectious needle due to accidental sliding.

0.45-mm needle corresponds to G26, marked by brown color coding. The term "gauge" is an old measure of thickness from early wire-drawer industry. Purely based on experiments, the sequence of sizes of tabulated gauge values is nonlinear, and different tables of gauge values have been coexisting. Eventually, in 1883, the British Standard Wire Gauge was proposed to replace the previously ill-defined gauge tables. In the algorithm defining the sequence of gauge sizes, values are separated by an integer multiple of approx. 0.004 inch (see Figure 3.21) to fit preexisting systems such as the so-called Birmingham Wire Gauge, but at the same time approximating the metric system, where 0.004 inch is close to 0.1 mm. In our days, despite seeming old-fashioned and not accurate, the gauge system has yet survived the standardization of measurement by introduction of S.I. units back in 1960. A major reason why the gauge system still exists is because it allows for quickly identifying needle thicknesses via simple numbers without need for checking many decimal places [22]. Along these lines, commercial suppliers (like Terumo, Braun, or Becton Dickinson) of hypodermic – meaning skin penetrating – needles, which are nothing more than hollow metal wires (= tubes), generally follow the British Standard Wire Gauge.

3.2.4 Optical inspection of microfluidic devices: Choice of microscopes and cameras

Once the microfluidic device assembly is completed and fluid reservoirs are connected, the next step is to identify a suitable observation setup onto which the microfluidic device can be mounted and operated. Due to the typical feature sizes of 1–100 µm in microfluidic devices, we typically require light or fluorescence microscopes for in-depth characterization of experiments. Here, we can distinguish between two types of microscopes: upright microscopes, where objectives are positioned above a static or x–y translational mechanical stage (see Figure 3.22A), and inverse microscopes, where objectives are positioned underneath the stage (see Figure 3.22B). When mounting a microfluidic device onto either microscope type, it is crucial to minimize the clearance and working distance, respectively, between the objective front lens and the glass substrate or the microchannels of the flow cell.[22] In this context, stamped microfluidic devices are more flexible in use than glass microcapillary-based devices as they can be mounted on both microscopes with upright and inverse light path. That is because glass capillary devices with their protruding needle-based fluid access ports (cf. Section 3.1.7) are challenging

---

[22] The objective's working distance decreases as the numerical aperture (NA) and the magnification increases. For example: for a standard simple achromat (achromatic lens system), the working distance decreases from 6 mm to 0.6 mm as the magnification increase from 10x at NA = 0.25 to 40x at NA = 0.65.

Color	Gauge[1]	Outer diameter (mm)	Outer diameter (inch)	Inner diameter (mm)	Inner diameter (inch)	Wall thickness (mm)	Wall thickness (inch)
ND	10*	3.40	0.134	2.69	0.106	0.36	0.014
ND	11	3.05	0.120	2.39	0.094	0.33	0.013
ND	12	2.77	0.109	2.16	0.085	0.31	0.012
ND	13	2.41	0.095	1.80	0.071	0.31	0.012
ND	14*	2.11	0.083	1.60	0.063	0.25	0.010
ND	15	1.83	0.072	1.37	0.054	0.23	0.009
	16 (1.60 mm)	1.65	0.065	1.19	0.047	0.23	0.009
ND	17*	1.47	0.058	1.07	0.042	0.20	0.008
	18 (1.20 mm)	1.27	0.050	0.84	0.033	0.22	0.009
	19 (1.10 mm)	1.07	0.042	0.69	0.027	0.19	0.008
	20 (0.90 mm)	0.91	0.036	0.60	0.024	0.15	0.006
	21 (0.80 mm)	0.82	0.032	0.51	0.020	0.15	0.006
	22 (0.70 mm)	0.72	0.028	0.41	0.016	0.15	0.006
	23 (0.60 mm)	0.64	0.025	0.34	0.013	0.16	0.006
	24 (0.55 mm)	0.57	0.022	0.31	0.012	0.13	0.005
	25 (0.50 mm)	0.51	0.020	0.26	0.010	0.13	0.005
	26 (0.45 mm)	0.46	0.018	0.26	0.010	0.10	0.004
	27 (0.40 mm)	0.41	0.016	0.21	0.008	0.10	0.004
	28 (0.36 mm)	0.36	0.014	0.18	0.007	0.09	0.004
	29 (0.33 mm)	0.34	0.013	0.18	0.007	0.08	0.003
	30 (0.30 mm)	0.31	0.012	0.16	0.006	0.08	0.003
	31 (0.25 mm)	0.26	0.010	0.12	0.005	0.06	0.003
	32 (0.23 mm)	0.23	0.009	0.11	0.004	0.06	0.003

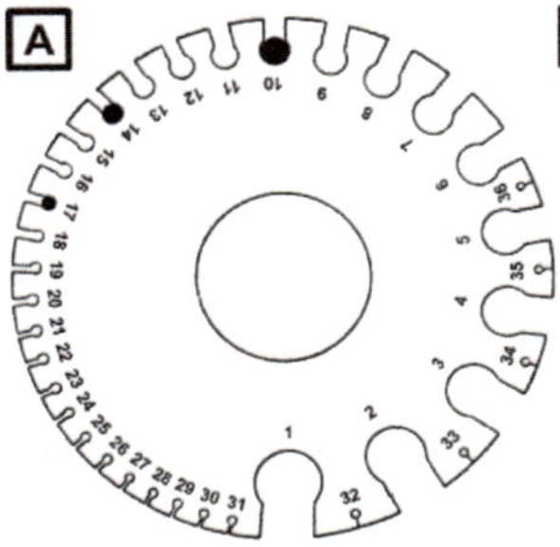

Figure 3.21: Different measures for the geometry of needles, which have potential relevance in microfluidics. (A) Wire gauge adapted from F. H. Rolt: "Gauges and Fine Measurements," London, Macmillan & Co., 1929. (B) Needles with different inner and outer diameter and wall thickness described by the gauge number in relation to the metric and Anglo-American system, and by typical color coding, with "ND" meaning "non-defined." The outer diameter is given according to the British Standard Wire Gauge from 1883. The gauge values of syringes highlighted with an asterisk are exemplarily related to the wire gauge in (A).

to be inserted into the lightpath of an upright microscope, and are thus usually operated on microscopes with inverse light paths. Stamped flow cells, by contrast, are made of elastomers like PDMS and merely require punch holes into which flexible tubing is plugged in. Thus, even when they are operated upside down, as shown in Figure 3.22B, the microfluidics setup is not too bulky to be inserted into the lightpath of an upright microscope. However, in this setup, connecting tubes face downwards toward the stage and are too flexible to provide a stable support for the flow cell. We thus suggest to put a spacer structure between the microflow cell and the microscope stage, as also shown in Figure 3.22B, which is basically a plate cut out in those areas where tubing is conntected to the above flow cell.

At this stage, the experimental setup for optimal optical inspection of microfluidic experiments is not yet complete, as we miss another key component. Flow pattern formation in microfluidic channels is fast; for example, in SF microfluidics, emulsion droplets can form at kilohertz rates, that is, 1,000 drops per second. The

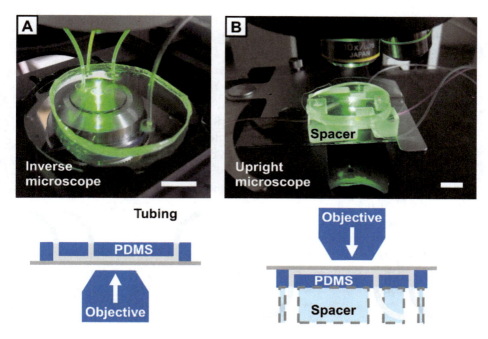

Figure 3.22: Mounting of a stamped microfluidic device onto two different kinds of microscope geometries. (A) Microflow cell installed on an inverse microscope with objectives underneath the flow cell, and (B) mounted onto an upright microscope with objectives positioned above the microscope stage. In the latter case, the flexible tubes that are connected to the flow cell cannot provide sufficient stability for in-plane characterization of the microfluidic experiments conducted in it, such that a solid spacer is put between the flow cell and the microscope stage with notches matching the position of the tube ports in the PDMS replica. The scale bars in both panels denote 1 cm.

human eye is merely able to resolve processes at an average rate of approx. 20 Hz; only some exceptional people may be able to detect the flickering of light or image artifacts in fast computer games at rates of up to 50 Hz. Thus, the maximum frame rate of conventional microscope cameras may not be sufficient to resolve fast microfluidic processes. Therefore, users new to the field of microfluidics may be told to equip their microfluidic experimental setup with fancy high-speed cameras with frame rates of 10,000 fps and more, coming along with significant investment costs of 10K to 50K Euros and more. However, even though microfluidics (SF microfluidics in particular) is usually associated with fast flow pattern formation, such ultrahighspeed cameras are not mandatory. Instead, standard USB3-connected digital cameras with fast shutter (meaning: exposure time in the range of just about 10 ms) and framerates of 50 to 100 fps are perfectly fine (cf. Section 3.2.4). In CF microfluidics, the task of imaging co-flowing liquids is not so hard anyways, as these flow patterns are usually time constant. And even in SF microfluidics, these cameras are fine, even though they cannot resolve each individual droplet formation if that occurs in the kilohertz range. This is because the periodic sampling of these cameras and the overlay of it with the (differently) periodic droplet formation leads to **strobing images**. These images seemingly show individual droplets that seemingly flow in weird patterns, such as flowing backwards, but even though this is an artifact from two superimposed differently periodic processes, these strobing images nevertheless allow users to identify if controlled dripping of uncontrolled jetting occurs in a device, and how large the momentarily forming droplets are.

In summary, for determining whether a desired flow pattern forms in a transparent microfluidic devices, a simple light microscope connected to a digital camera with up to 1,000 fps is sufficient for both CF and SF microfluidics. Cameras with USB3 connection can be obtained for less than 1,000 Euros, a light microscope equipped with two achromat objectives (10x, 40x), a stationary microscope table,[23] and long-lasting LED light source[24] is available for 2,000 to 5,000 Euros on the European market, for instance.

To this point, the characterization of microfluidic devices has been based on flow cell imaging by simple bright-field microscopy. Yet, for in-depth analysis of the flow inside a microfluidic device, we require characterization methods that allow for determining velocity profiles and quantitative mapping of concentration profiles in microchannels; for example, for quantifying the distribution of reagents at low concentrations – both small organic molecules (such as dyes) and (bio-)macromolecules (such as DNA or enzymes) – or the degree of mixing of reaction partners to determine the effectiveness of a micromixer. Here, we want to highlight two techniques that are most likely available at a reader's research institute and pose the state of the art in microflow imaging, which is confocal laser scanning microscopy (CLSM) or confocal spinning disc microscopy (CSDM),[25] and microparticle

---

[23] Translational x–y microscope tables with clamps as sample holders are nice to have, but a cost factor. Instead, an experimenter may align the microfluidic device in the light path of a microscope and then fix the microfluidic device with tape that can be easily removed and re-attached to the microscope table (e.g., Scotch Tape, 3M).

[24] A great advantage of LEDs over conventional light bulbs is that LEDs usually do not heat up as much as light bulbs and are thus more resistant for use in cold rooms at low temperature and high humidity or for use in experiments with temperature-sensitive fluids.

[25] A major disadvantage of conventional CLSM is that it is based on moving a spot of light emanating from a single pinhole over a sample. Image acquisition is thus rather slow, requiring extended illumination for analyzing large-area and large-volume samples, which may lead to high levels of bleaching. While 3D reconstruction of a fluorescent flow profile is usually not impaired by photobleaching, as the fluorescent species of interest is renewed every microsecond in the continuous flow, CLSM may be too slow to resolve fast processes at the interface of liquid–liquid streams. To account for this issue (and potential photobleaching), microfluidics users may consider to apply CSDM for flow profile analysis. As CSDM generates multiple beams from a spinning pinhole disc (so-called Nipkow disc) matched in position and speed with a second microlens disc, 3D imaging can be performed at high frame rates (approx. 1,000 fps) with just little photobleaching and thus extended acquisition time, if necessary.

image velocimetry (µPIV) [23]. These characterization methods allow for tracing fluorescent marker molecules or particles, whose displacement in 2D and even 3D yields information on the degree of mixing for a fluorescent species with defined molecular weight and size (confocal microscopy) or the velocity profile within a microchannel by the residence time of microparticles with defined size within a distinct microchannel length (µPIV). For instance, due to the laminar flow conditions at low Reynolds numbers in CF microfluidics, concentration profiles in a flow-focusing experiment with two (or more) miscible liquids are stationary (cf. Section 3.3). That means, if we walk along the outflow channel of a CF microfluidic experiment, the local composition at the co-flowing liquid–liquid interface that forms at the microchannel junction and extends into the outflow channel is constant. This very stability of flow patterns in CF microfluidics allows for utilizing confocal microscopy methods to not only visualize fluid jets in a 3D fashion, but to also image diffusion processes at its liquid–liquid interface, as shown in Figure 3.23, and to further correlate these results with FEM simulations (cf. Section 1.1.4).

3.2.5 Starting microfluidic experiments

With a microflow cell and an experimental setup at hand, we now come to a delicate point: start-up of the microfluidic experiment. Obviously, flow rates need to be chosen adequate to a desired flow pattern (such as well-defined laminar co-flow of desired relative extent in CF microfluidics, or controlled dripping in SF microfluidics); flow rates also need to be chosen such to prevent the microfluidic device from delamination due to excess pressure. But how do we find these rates, and how do we start at all from an initial stage of zero flow? For a profound choice of starting flow rates, one should consider some fluid and microchannel parameters, most importantly, the fluid viscosities and compressibilities as well as the microchannel surface wettability. For experimenters new to the field of microfluidics, it is usually easier to identify suitable starting flow rates for CF microfluidics, as these "only" involve the formation of a stable, non-transient interface of two miscible fluids.[26] In case of a Y-junction, the flow rates may be optimized to focus the liquid–liquid interface in the center of the outflow channel, while for a cross-junction with three inflow channels as utilized in flow-focusing experiments, the flow rates may be optimized to achieve a certain fluid jet width and position within the microchannel. For SF microfluidics, the flow physics of the droplet formation mechanism as well as their dynamic, moving interface formed by nonmiscible fluids hamper identifying suitable starting flow rates.

---

[26] Nevertheless, a pair of miscible fluids may need to be formed from strongly dissimilative materials, e.g., from water and a high-molar-mass polymer solution. Then, adequate flow rates are required to prevent backflow of the viscous polymer solution into the water inflow channel, for instance.

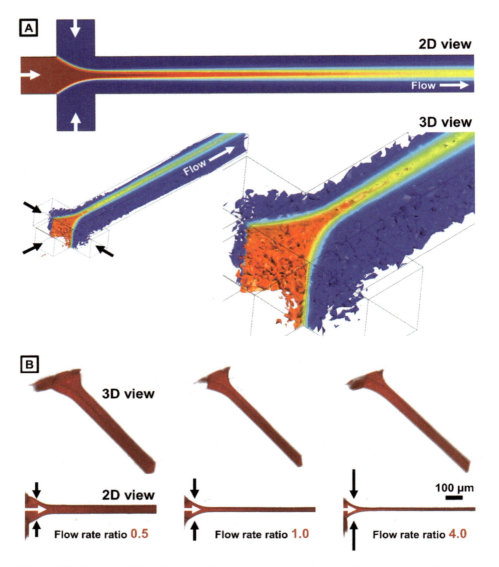

Figure 3.23: Examples of detailed simulation and characterization of mixing in a microfluidic flow-focusing experiment. (A) FEM simulation of (B) 3D reconstruction of a stack of confocal laser scanning microscopy images.

Why are the initial flow rates of such significance? The reason is simple: microchannel properties as well as the physicochemical fluid properties define the inflow speed, and with that, the time it takes for a fluid to reach a flow cell's key microchannel feature. As we will again discuss in Section 3.5, it is preferable to have all required fluids in place at the same time, for example, at the center of a cross-junction. That way, no compressible air remains in any of the inflow channels that may promote fluid **backflow**. Fluid backflow describes the transport of fluid from one inflow channel into another inflow channel instead of the designated outflow channel. For miscible, nonreactive fluids, backflow can be reset by simply increasing the backflowing-fluid's flow rate to drive it out. The use of immiscible fluids (oil and water, for instance) may aggravate the effect of fluid backflow, depending on the device design (cf. Section 3.2). For instance, backflow of water into the oil inflow channel and vice versa may lead to local uncontrolled emulsion formation at the oil inflow port, which disturbs the constant fluid flow and impairs the uniformity and monodispersity of the emulsion product in the initial stage of the microfluidic experiment. That accidental emulsion formation is particularly severe if the inflow ports of the microfluidic device contain filter units (cf. Figure 3.3B), as these act like a sieve and promote uncontrolled droplet formation in the case of backflow. Even worse, polydisperse emulsion droplets that are accidently formed in these filter units may get trapped there, and then, at a later stage of the experiment, get loose and flush into the emulsion that then steadily forms downstream in its actual site of formation, the channel cross-junction, thereby impairing the overall emulsion uniformity again. So, when starting a microfluidic experiment, we recommend to thoroughly monitor all incoming fluids and regulate their flow rates such to have them meeting each other at the same time in the designated key site of the microchannel system, which most typically is a channel junction. To save time, it is helpful to advance the fluids to the end of their respective inlet tubing by high flow rates first, and then, once they enter the microfluidic chip, to drastically lower the flow rates such to have the fluids entering the channels and advancing forward to the junction of interest at low speed, manually fine-tuned further such to have them all meeting each other at the same time.

Beyond the very first start-up of a microfluidic device, flow rates need to be identified that allow for desired flow pattern formation in CF and SF microfluidics. Again, finding a suitable set of flow rates for CF formation is generally easier compared to finding a suitable set of flow rates for droplet formation, as droplets can be formed in different modes of fluid flow, for example, in a geometrically controlled dripping regime or in a less controlled jetting regime (cf. Chapter 2). As a rule of thumb, for Newtonian, incompressible fluids exhibiting water-like viscosities (kinematic viscosity of water: $v = 0.8926$ mm^2 s^{-1};[27] kinematic viscosity of fluorinated oil, for example, HFE7500: $v = 0.77$ mm^2 s^{-1}), a **flow-rate ratio** of approx. 1:3 to 1:10 (dispersed aqueous phase: continuous oil phase) is satisfactory for droplet formation in a microfluidic cross-junction. For instance, assuming microchannel dimensions of $h = w = 50$ μm, the aqueous phase can be set to 200 μL h^{-1} and the oil phase to 600 μL h^{-1} to achieve controlled droplet formation. Note: this initial set of flow rates may be further optimized depending on the desired droplet volume and

---

[27] v (mm^2 s^{-1}) = v (cSt).

shape (e.g., when passing through a narrow microchannel) as well as the interdroplet distance in the outflow channel along the microfluidic experiment.

On top of the above reasoning, it is not sufficient to only operate a flow cell at a correct flow-rate ratio, but also at correct absolute rates. For instance, the above discussed 50 μm-cross-junction could also be operated at flow rates of IP = 2,000 μL h^{-1} and CP = 6,000 μL h^{-1}, which would still satisfy the previous statement that the flow-rate ratio of the dispersed and the continuous phase should be in a range of approx. 1:3–1:10 for droplet formation. Yet, an experimenter using this set of flow rates will most likely observe that the flow cell will not form microdroplets, but a continuous jet extending from the cross-junction far into the outflow channel. Here, we have to recall the delicate interplay of shear and pressure effects leading toward either dripping or jetting at microchannel cross-junctions, as previously discussed in Section 1.5.3 (cf. Weber number We and Capillary number Ca). Further details on identifying correct flow rates (ratio and overall rate) as well as recognizing flow conditions in a running microfluidic device are discussed in Section 3.5.

To this point, we have solely focused on correctly starting up a microfluidic device, but it is also important to look further downstream at the outflow channel and the attached collection reservoir. Microfluidic experiments may last from minutes to days. If we leave a product like an emulsion or suspension in a collection reservoir in free air, it will quickly evaporate due to its large surface-to-volume ratio. For instance, if just-prepared emulsion droplets that are meant to serve as templates for polymer particle preparation (cf. Section 3.4.2) are left in an Eppendorf tube without covering the open collection container, solvent evaporation from the emulsion will lead to an increase of material concentration inside each droplet and thus directly influence material properties of the polymer particles regarding their degree of crosslinking and polymer-network nanoscopic topology [24]. To slow down the solvent evaporation from the collected microfluidic product, the liquid sample should be covered by light mineral oil in its collection vessel, which exhibits a lower density than water ($\rho \approx 0.84$ g mL^{-1}) and is chemically rather inert. Alternatively, one may stretch a piece of Parafilm® tape (a foil with low gas permeability made from a mixture of paraffin wax and polyethylene) across the open collection vessel and punch the outflow tubing through the tape. As the tubing is inserted into the collection vessel, one should take care that the tubing has been cut flat and does to touch the bottom of the collection vessel. Both means will ensure that the microfluidic product (e.g., droplets) exiting the tubing is not squeezed between the tubing tip and the bottom of the collection vial or ripped apart at sharp edges of the tubing.

Before permanently connecting the microfluidic device to a collection reservoir, the experimenter should make sure that the desired microfluidic products are not only formed in a monodisperse fashion, but that they also retain their material properties as they exit the device. For this purpose, the experimenter should take a fluid sample directly at the outflow port at the beginning of every microfluidic experiment to check its quality, as PDMS debris and dust from the device fabrication process may still reside inside the outflow port – even if the microchannel replica has been washed extensively before – and disrupt delicate products like droplets and vesicles.

3.2.6 Running microfluidic experiments

A steady inflow of fluids at stable flow rates should enable an experimenter to leave a microfluidic setup running without constant monitoring of its performance. Yet, the microchannel surface properties can deteriorate due to *adsorption* of molecules at the surface of the microchannels as well as due to *absorption* of solutes into the microchannel walls – particularly in case of microchannels molded into PDMS (as discussed further in Section 3.5). Furthermore, particulate or undissolved content may clog the flow cell over time, especially when dead zones of fluid flow are not sufficiently accounted for. Moreover, a certain set of flow rates may turn out to only give kind of a meta-stable regime of device operation, which may switch to an undesired state (such as jetting in SF microfluidics) just "out of nothing," for example, in consequence of temporary small fluctuations of the fluid flow rates. The device may even jump back and forth between two regimes of operation of flow rates are chosen at the border between these two regimes. Such an unexpected jump from droplet formation by controlled dripping into less controlled jetting may not only be caused by a meta-stable flow regime, but also by a surface wetting defect of the outflow channel (Section 3.5). In planar microflow cells, where the microchannel network solely extends in the $x–y$ directions, the inner dispersed phase is in contact with the upper and lower microchannel walls in the moment of droplet formation. Thus, if the wettability of the outflow microchannel is insufficient or patchy and thus not necessarily favoring wetting of the continuous phase, the inner phase will form a jet on the surface of the outflow channel extending into that channel without droplet formation. If a device fell into such a state, simply ramping up the outer-phase flow rate by resetting the pumping system to reinitiate droplet formation may not be sufficient. The inertia stored in the flow cell as well as the tubing connection between the device and the pumping system (cf. Section 3.2.3) will prevent a sudden change of the flow rates, leaving the inner phase connected to the microchannel walls. Although it may sound unprofessional, but a sudden change of the flow conditions inside a microfluidic device can be achieved by simply snapping the tubing that transports the continuous phase. The induced fast-traveling flow front can detach the inner phase from the microchannel wall and reinitiate desired droplet formation in the outflow channel despite a microchannel wetting defect.

To account for the afore-listed sources of microfluidic device failure during continuous operation, the experimenter may consider automated observation of the microfluidic experiment, for example, by periodic image acquisition. This way, potential variations of the microfluidic product quality (emulsion size distribution, encapsulation efficiency) can be related to any event occurring during the product formation, such to rule out that source of error in the future.

If we assume that the microfluidic device is running under stable flow conditions, we can move to the next challenge of microfluidics, which is the low product output of microflow cells due to fluid miniaturization (cf. Section 2.5). Due to that, the product output is usually in the range of just a few micro- to milliliters per hour. As fluids are employed that are rather easy to handle (oils- and water-based liquids, mostly dilute solutions with Newtonian flow), microfluidic experiments may be performed overnight without continuous observation as long as the flow pattern is stable and device degradation is monitored (cf. above). Note: if a microfluidic device is coupled to a light or fluorescence microscope, the experimenter should account for the heat input into the microfluidic device by the focused light in the microscope beam bath, particularly when processing temperature-sensitive fluids for an extended period. If image-based surveillance is desired to monitor long-term microfluidic experiments, the device illumination should therefore be ideally coupled to the actual image acquisition and not just performed continuously throughout the microfluidic experiment.[28]

3.2.7 Finishing microfluidic experiments

Once the fluid reservoirs connected to a running microfluidic device are running low, it is important to take measures that provide consistent product quality and minimize its loss near the end of a microfluidic experiment. In case of (single-)emulsion formation, for instance, one should keep the continuous phase running after the inner or dispersed phase has been emptied to ensure that the desired emulsion droplets that are still in the outflow channel and tubing are flushed out into the collection reservoir. For that, the experimenter should not only make a rough estimation of the fluid consumption at the beginning of the microfluidic experiment with regard to the expected flow-rate ratios and have the reservoirs filled up accordingly (cf. Section 3.2.5), but also account for an extra volume of the continuous phase to flush the microflow cell at the end of the experiment. If experimenters cannot be present toward the end of a microfluidic experiment, it is crucial to take additional measures depending on the pumping system that is in use to ensure successful completion of the experiment. While in pressure-driven pumping systems, the flow will just stop as soon as the pressurized reservoir is emptied without harming the pumping system, it is absolutely necessary to program a "stop" command into conventional positive-displacement syringe pumps that stop the propulsion of the syringe piston. Otherwise, the screw-drive of the pump will continue to push the piston against the – now empty – syringe body, and ultimately break it, which – in case of glass syringes – can be a rather expensive event (e.g., a Hamilton GASTIGHT® 1001 TLL syringe costs approx. 70 € in 2018).

---

28 which, in parallel, increases the light source's lifetime.

As the microfluidic experiment finishes, it is also – plainly spoken – "pay day," and the experimenter can finally reap the fruit of one's work. For example, if we collect the outflow of a microfluidic device containing a 50 μm cross-junction for single-emulsion formation, being operated at 200 μL h^{-1} for the dispersed phase and 600 μL h^{-1} for the continuous phase for approx. 10 h, we will find as much as 8 mL of emulsion in the collection vessel. While this yield cannot compete with industrial standards, it is perfectly fine for further characterization or utilization of the microfluidic product in academia – for example, as a template for further material design (microgels or vesicles) – as one has to recall that the typical frequency of droplet formation is in the hertz-to-kilohertz range, such that 8 mL of emulsion may contain as many as 30 billion droplets!

After finishing a microfluidic experiment, users may desire to reuse a microfluidic device for further experiments to avoid repeated elaborate device replication by soft lithography. In that case, it is crucial to remove any solutes, particles, or the like from the microfluidic device, as all of these species may adhere to the microchannels, the stronger the longer they reside inside the flow cell. Particularly, in SF microfluidics, where droplet formation commonly requires the use of surface-active molecules, exactly those permanently adhere to the microchannel surface in form of a thin layer after solvent evaporation during flow cell storage. This layer of amphiphiles may invert the wettability pattern of the microflow cell, thus causing formation of a correct flow pattern in a follow-up experiment (cf. the following sticky notes).

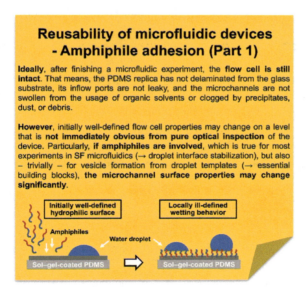

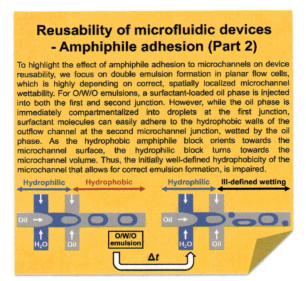

3.3 The practice of continuous-flow microfluidics

3.3.1 Single-phase CF microfluidics: Analytics, sensing, synthesis

The power of CF microfluidics lies in its ability to provide quasi-static, highly controlled experimental conditions. Two major fields of application of these controlled conditions are (bio)chemical synthesis and (bio)chemical analytics. Here, by utilizing laminar flow and minimal sample consumption, microflow cells provide controlled fluid–fluid contact and thereby allow for performing reactions with superior spatiotemporal resolution in solution or on microchannel surfaces, which can then be analyzed. Even without use of multiphase continuous flow, a smart combination of microchannel architecture and fluid choice can already provide multiple purposes even in single-phase flow. For instance, we can utilize microchannel networks to mimic key elements of solid-state electronics, as detailed in Section 3.1. Here, fluid flow can be manipulated similarly to electron flow by microchannel features that take over the function of resistors, capacitors, and oscillators. For instance, a key element in microprocessors is the flip–flop memory responsible for storing the state of information "0" or "1," which is one bit. The realization of such switchable elements in CF microfluidics is based on utilizing liquids that show nonlinear flow phenomena at low Re numbers, for example, non-Newtonian, viscoelastic polymer solutions. The corresponding flow cell design is simple and merely relies on a Y-shaped microchannel in which a metastable fluid stream of a high-molar-mass polyacrylamide or poly(ethylene glycol) solution, for instance, is guided toward one of two outlet ports, corresponding to the states "0" and "1." In contrast, Newtonian fluids would evenly split at the Y-shaped separator without control over the switching behavior of the microfluidics-based microelectronics mimic [25].

Beyond just *mimicking* microelectronic circuits and devices by CF microfluidics, *integration* of microelectronic components into flow cells is likewise feasible. Exemplarily, this integration is frequently performed in **microfluidic sensors** (cf. Figure 3.24). Here, due to their free design choices for microchannel networks and their capability of miniaturization and integration of multiple functions (e.g., liquid pumping, storing, and manipulation), microfluidic devices are promising experimental platforms for performing in-field analytics in disposable lab-on-a-chips with integrated transducers. A key application of these **integrated devices** lies in the field of fast food testing, for example, peanut allergen detection. In a typical flow cell layout, a microchip-based mass sensor is positioned underneath a microchannel in which the analytes of interest are passed through by integrated capillary pumps (cf. Figure 3.24). Changes in the vibration frequency of the mass sensor membrane (Figure 3.24) then allow for quantifying the content of analytes bound to the sensor surface, and thus in the food sample. A key challenge in that type of devices lies in their leakage-free assembly when interfacing solid microchips with the elastomer-based microchannel replica. Here, it is advised to generate a height profile of the to-be-integrated microchip first, for example, by confocal microscopy (cf. Figure 3.24A), which is then translated into a CAD file to fabricate a multilayered negative profile by photomasks-based photolithography. This master device is then replicated by PDMS-based soft lithography to yield a flow cell that matches exactly the structure of the microchip.

In addition to simple passive single-phase fluid flow, a further variant of CF microfluidics is to actively *modify* the flow of fluids during their analytics. With such active modification, CF microfluidics provides means to study the analytes' mechanics, composition, and morphologies, including that of living cells and other objects such as nano- or microparticles, micelles, and vesicles, in a continuous high-throughput fashion. At tailored fluid flow and microchannel dimensions, for example, orientation or even deformation of analytes can be achieved due to shear forces. These forces originate from the no-slip boundary condition at the microchannel walls, leading to a parabolic flow profile that causes deformation of soft particulate content (e.g., cells) inside the flow. Alternatively, microchannel constrictions can be utilized to generate even stronger shear forces onto particulate fluid flow. For instance, by utilizing microchannel constrictions in single-phase flow experiments or narrow fluid jets generated in two- or three-phase flow experiments (cf. below), cells can be forced into a single-file flow and analyzed one after another, which allows for investigating their properties (for example, their mechanics) in a high-throughput fashion instead of studying them in conventional multi-cell cultures. Another example of application is to force colloidal-scale cylindrical micelles through a microchannel tapering to study their alignment within versus behind a constriction, as shown in Figure 3.25.

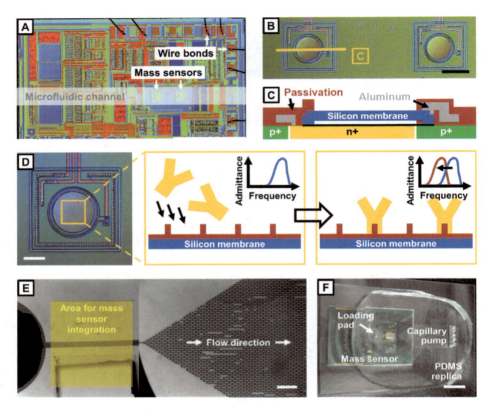

Figure 3.24: Food allergen detection via antigen–antibody reactions on mass-sensitive vibrating membranes. (A) Differential interference contrast (DIC) microscopy analysis of a microchip with integrated mass sensors based on vibrating silicon membranes. Wire bonds provide connection to an external power supply and read-out or control. The desired alignment of the sensors with a microchannel feeding the sensors with liquid food samples is highlighted. (B) Zoom of a section of the microchip from (A) showing the mass sensors. One mass sensor remains unfuctionalized and thus does not allow for any antigen–antibody reaction but instead serves as internal reference. The scale bar is 100 μm. (C) Schematic cross section of a mass sensor. (D) Schematic of the antigen–antibody reaction causing a shift in the vibration frequency. The scale bar in the left image denotes 50 μm. (E) Bright-field microscopy image of a microflow cell flushed with an aqueous food dye (Brilliant Black BN, 0.01 M) solution for enhanced contrast. The designated area for the mass sensor microchip beneath the straight microchannel is highlighted. The scale bar denotes 1 mm. (F) Mass sensor integrated in a PDMS-based microflow cell on a glass substrate. The PDMS replica contains a capillary pump in the right part of the device, and a round loading pad for liquid samples in the left part above the mass sensor. The scale bar is 1 cm.

The manipulation of colloids that continuously flow through microchannels is not limited to a certain material basis (like living cells or organic cylindrical micelles, as just discussed), but also applicable to suspensions of inorganic colloids such as nanowires. Here, depending on their rigidity and length in relation to the

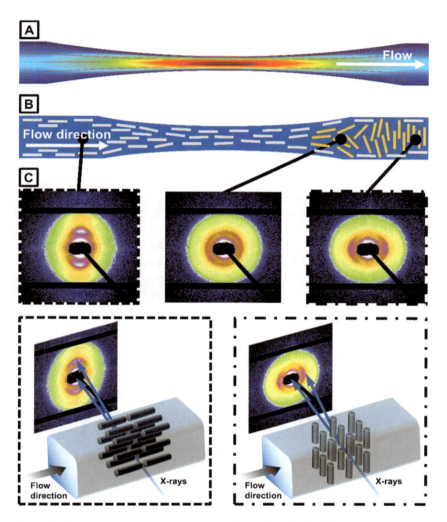

Figure 3.25: Analysis of a suspension of cylindrically shaped colloidal-scale objects such as cylindrical amphiphilic or polymeric micelles continuously flowing through a microchannel constriction by high-resolution, micro-focus small-angle X-ray scattering (SAXS). The characterization is supported by polarization microscopy and velocity field simulation [15]. The cylindrical analytes are well-aligned parallel to the flow before and within the microchannel constriction. However, counter-intuitively, increase of the flow speed within the tapering does not lead to further improvement of the alignment behind the tapering. Instead, the cylindrical analytes flip their orientation by 90°, as also observed for other anisotropic particles including worm-like surfactants and disk-like particles, indicating the generality of this phenomenon. Realignment is caused by the extensional flow perpendicular to the flow direction in the expanding microchannel. This is promoted by shear thinning, a common property of polymer solutions. These experiments also provide valuable insights in established production processes – plastics extrusion, for instance – where reorientation of polymer fibers in the extrusion direction may be required to improve their mechanical performance. Partly reproduced from [15]; copyright 2013 National Academy of Sciences.

microchannel dimensions, these highly anisotropic particles can be trapped in arrays of parallelized microchannel constrictions on a silicon wafer, which is then contacted by UV lithography to yield multiple nanowire-based field effect transistors in a parallelized fashion [26].

3.3.2 Multiphase CF Microfluidics: Mixing, gradients, synthesis, nucleation, self-assembly, surface patterning

Through the use of Y- or X-shaped microchannel junctions, multiple flowing streams can be combined and co-flown alongside each other in a laminar fashion in CF microfluidics; this kind of flow pattern has enormous potential and utility in bio-analytics and materials design, for example, for studying fast reaction kinetics of protein folding, early stages of nanoparticle growth, sudden phase transitions in complex polymer materials, and vesicle formation. Furthermore, fast homogenization of co-flowing reaction partners, and with that, control of their reaction onset, allows products with narrow size distribution to be obtained, such as in the case of polymerization reactions, particle nucleation, or colloidal self-assembly, as summarized in Figure 3.26.

The most useful variant of fluid interfacing in multiphase CF microfluidics is **hydrodynamic flow-focusing** (HFF), as sketched in all examples in Figure 3.26. In this variant, two or three fluids meet at a four-way cross-junction such that the one

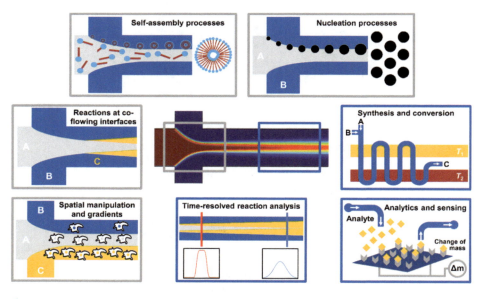

Figure 3.26: Overview over key applications of CF microfluidics based on co-flowing interfaces at microfluidic cross-junctions (gray frames), leading to either single- or multiphase flow downstream the outflow channel (blue frames).

entering centrally is focused down to a narrow stream by the two entering perpendicularly. HFF experiments provide access to reaction kinetics and out-of-equilibrium process dynamics that cannot be resolved in conventional macroscopic mixers, for instance by stirring. This is because the time evolution of a reaction in a CF microfluidic channel scales in proportion to the distance from the location of the very first contact of the fluids. Due to the inherently low Reynolds number flow in microfluidics, the concentration profile inside such a flow-focused fluid jet is determined by diffusion only, and can therefore be precisely estimated at any point of the reaction. As a result, a precise point in the reaction progress can be assigned to an individual position inside the microchannel, as illustrated in Figure 3.27. This way, (bio-)chemical reactions with extremely fast kinetics can be resolved, and colloids (nanoparticles and vesicles) can be prepared with defined sizes by controlling the flow-rate ratio and the overall flow rate in the HFF experiment. Figure 3.28 visualizes a realization

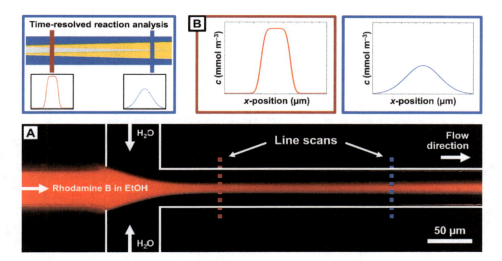

Figure 3.27: CF microfluidics for spatiotemporal investigation of reaction kinetics and diffusive mixing at liquid–liquid boundaries. (A) Fluorescence micrograph of a microchannel cross-junction with three inflow channels, realizing a hydrodynamic flow-focusing experiment to study the diffusive mixing of the inflowing components at their liquid–liquid interfaces. The central inlet is fed with a solution of the fluorescent dye rhodamine B in ethanol, whereas the two side inlets are fed with water. By adjusting the flow-rate ratio of the central channel inflow and the two side channel inflows, a focused liquid jet with a diameter as small as 1 µm can be formed. (B) Due to the laminar flow conditions, the diffusive mixing at the liquid jet's interface can be studied with high spatiotemporal resolution, because each position along the outflow channel corresponds to a defined time point of the mixing process. In the experiment shown here, two line-scans of the fluorescence intensity are performed perpendicular to the flow direction at positions indicated by the red and blue dotted lines. These scans yield intensity profiles as sketched in the red- and blue-framed schemes above; they quantify the lateral diffusive smearing of the flow-focused rhodamine jet downstream the outflow channel.

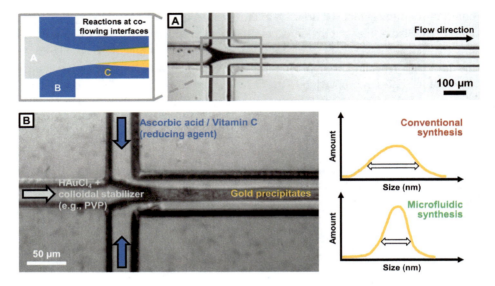

Figure 3.28: Spatiotemporal investigation and control of (bio-)chemical reactions at co-flowing interfaces utilizing CF microfluidics. Exemplarily, the synthesis of gold nanoparticles at room temperature is shown. (A) Hydrodynamic flow-focusing in a PDMS-based microfluidic device. Depending on the flow-rate ratio, the volume ratio of reaction partners in a given fluid volume in the outflow channel can be adjusted, and with that the reaction outcome, for example, with regard to the product particle size. (B) Preparation of gold nanoparticles by a redox reaction in a flow-focusing experiment. Compared to conventional gold nanoparticle synthesis in macroscopic bulk solution (e.g., in a reaction flask), fast and efficient mixing at the liquid–liquid boundary of a gold salt solution and a reducing agent allows for controlling the onset of particle nucleation and thus for obtaining a narrow particle size distribution.

of the latter example: the formation of gold nanoparticles in a CF microfluidic experiment. On top of that, as microfluidic HFF requires just small volumes, precious biomacromolecules (such as DNA or genetically optimized enzymes) can be studied under mechanical stress utilizing the influence of geometric constraints on the flow at the interface of the flow-focused liquids, thereby extending the manipulation of materials as previously discussed in Section 3.3.1.

The distinct and controlled concentration gradients at the liquid–liquid contact zone in laminar CF microfluidics can also direct the formation of materials by nucleation and growth. A well-established class of application is microfluidics-assisted **nanoprecipitation**. In this technique, a polymer or (block-)copolymer solution that flows in a microchannel is subject to diffusive penetration of a non-solvent upon laminar co-flow; as a result, the polymer precipitates in the interdiffusive mixing zone. In general, nanoparticle formation by polymer precipitation occurs as a sequence of nucleation and growth of aggregates of macromolecules, which stops when colloidal stability is reached. In microfluidics, this process is tunable by the fluid intermixing rate and spatiotemporal profile, which is in turn controlled by the fluidic flow rates. A popular material that has shown promising performance in view of both its ability to form nanoparticles and its prospect to host drugs in actual biomedical applications is poly(lactic acid) and its copolymers with glycolic acid and ethylene oxide, as shown in Figure 3.29. In this example, control over the particle size is exerted by the initial polymer concentration in an organic solvent stream (here, acetone) and the liquid jet width, which is to be set by the fluid flow rates.

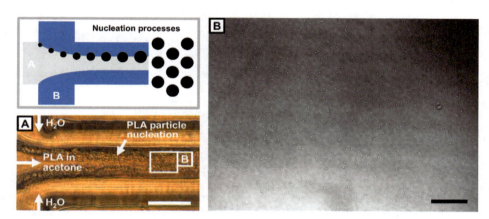

Figure 3.29: CF microfluidic experiment for studying and steering nucleation processes using the example of poly(lactid acid) (PLA) microparticle formation. (A) A solution of PLA (1%, w/w) in acetone as a good solvent for PLA (100 μL h^{-1}) is flow-focused into a liquid jet by water (400 μL h^{-1}) in which PLA is insoluble. Diffusive mixing at the liquid–liquid boundary induces PLA particle nucleation. Due to laminar flow conditions, each state of particle formation corresponds to a position along the outflow channel. Thus, PLA particles may agglomerate along the liquid jet's interface and eventually clog the microchannel over time (cf. Section 3.5). The scale bar is 25 μm. (B) Bright-field microscopy image of PLA microparticles dispersed in water, as collected at the outflow channel. The scale bar denotes 50 μm.

A peculiar variant of the nanoprecipitation method is the use of microfluidic environments to foster the inherent tendency for environmentally sensitive assembly of block copolymers such to forming vesicles. For example, microfluidic devices have been used for the directed assembly of unilamellar poly-2-vinylpyridine-b-poly(ethylene oxide) (P2VP-PEO) vesicles named **polymersomes** by flow-focusing of an ethanolic block copolymer solution in a stream of water, as shown in Figure 3.30. Alteration of the flow-rate ratio allows the vesicle size to be tuned over a range from 40 nm to 2 μm.

In contrast to microfluidics, conventional preparation methods of lipid- or polymer-based vesicles (e.g., film rehydration, electroformation) provide just limited control over the resulting size distribution, and each technique is limited to a certain size range. CF microfluidics, by contrast, enables the vesicle size to be tuned over several orders of magnitude (here, 40 nm to more than 2 μm) in the same

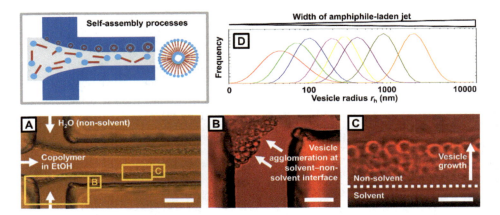

Figure 3.30: CF microfluidic experiment for studying and controlling self-assembly processes at continuously flowing liquid–liquid interfaces using the example of vesicle formation from natural (e.g., lipids) or synthetic (e.g., copolymer) building blocks. (A) Bright-field microscopy image of a microfluidic flow-focusing experiment. The vesicle-forming copolymer poly(2-vinylpyridine)-*block*-poly(ethylene oxide) (P2VP-PEO) is dissolved in ethanol (0.05%, w/w) and focused into a jet (50 μL h^{-1}) by plain water (100 μL h^{-1}). The block-length ratio is optimized to promote spontaneous self-assembly of the copolymers into vesicles – which are termed polymersomes in the case of copolymer building blocks – in aqueous solution. For forming vesicles by CF microfluidics, it is essential to use a set of miscible liquids, where the jet is made of a good solvent for the vesicle building blocks, and the jet-forming, surrounding liquids are a poor solvent for the building blocks. The scale bar denotes 10 μm. (B) As the solvent quality for the P2VP block drops at the liquid–liquid interface due to diffusive mixing, copolymer aggregates and eventually vesicles form at the jet's interface. The scale bar is 25 μm. (C) Over time, vesicles adhere to the microchannel surface. The hydrophobic vesicle membrane is fluorescently labeled by addition of the hydrophobic dye Nile Red to the initial copolymer solution. (D) Dynamic light scattering (DLS) experiments allow for determining the size of the vesicles as received by CF microfluidics. This size can be adjusted from approx. 40 nm to several micrometers by controlling the flow-rate ratio (FRR) of the copolymer jet and the surrounding flow-focusing phase. The scale bar is 5 μm.

experiment and avoids post-processing steps to manipulate the vesicle shell characteristics and size. The exact control of vesicle formation by HFF is based on the flow-rate ratio of the co-flowing amphiphilic liquids or polymers, which determines the width of the amphiphile-loaded center stream (see Figure 3.30D) and, to a less extent, the shear stress applied to the amphiphiles at the liquid–liquid interface. In a simple nucleation-and-growth model that is applicable to both liposome and polymersome formation, vesicle nuclei form at the interface of a flow-focused jet and grow by further uptake of vesicle building blocks (lipids or synthetic amphiphiles like block copolymers). Since the amount of amphiphiles is proportional to the width of the flow-focused liquid jet in a given microchannel volume element, large vesicles self-assemble from flow-focused jets with large width, and vice versa, small vesicles self-assemble from focused streams with small jet width. There is just one disadvantage of HFF-based vesicle preparation: with that method, the encapsulation efficiency of cargo payloads is hardly controllable. For quantitative encapsulation, thus, a different microfluidic technique is superior: utilization of double-emulsion templates, as detailed further in Section 3.4.5.

Note: As general as our so-far discussion on vesicle formation sounds, there is a restriction: not all amphiphiles can self-assemble into complex 3D structures, as certain prerequisites for the composition and structure of the amphiphiles exist. To quickly assess whether an amphiphile such as a block copolymer is potentially able to form vesicles, we need to consider the dimensionless packing parameter, P, which expresses the molecular shape of copolymers in solution, and thus the morphology of the corresponding self-assembled copolymer aggregate upon phase separation of the hydrophobic and hydrophilic blocks. This parameter is defined as the size of the hydrophobic block relative to that of the hydrophilic moiety.

$$P = \frac{v}{a \cdot l} \quad (3.1)$$

Here, v denotes the volume of the hydrophobic block, a describes the hydrophilic–hydrophobic interfacial area, and l is the hydrophobic block length normal to the interface. As P increases, the vesicle morphology is tuned from spherical over toroidal to cylindrical, as exemplarily shown in Table 3.2.

Table 3.2: Packing parameter P of different amphiphiles and corresponding values for the Gaussian curvature K and mean curvature H of their assemblies, which are expressed by the two radii of curvature r_1 and r_2.

Shape	$P = \frac{v}{a \cdot l}$	r_1	r_2	H	K
Sphere	$\frac{1}{3}$	r	r	$\frac{1}{r}$	$\frac{1}{r^2}$
Cylinder	$\frac{1}{2}$	r	∞	$\frac{1}{2r}$	0
Bilayer	1	∞	∞	0	0

As the vesicle shape is mainly determined by the interfacial curvature, the packing parameter P can likewise be described by the Gaussian curvature K and the mean curvature H with the two radii of curvature r_1 and r_2, as illustrated in Figure 3.31.

Formula 3.1 can thus be expressed as:

$$P = 1 - Hl + \frac{Kl^2}{3} \Rightarrow P = 1 - \frac{l}{2}\left(\frac{1}{r_1} + \frac{1}{r_2}\right) + \frac{l^2}{3r_1 \cdot r_2} \quad (3.2)$$

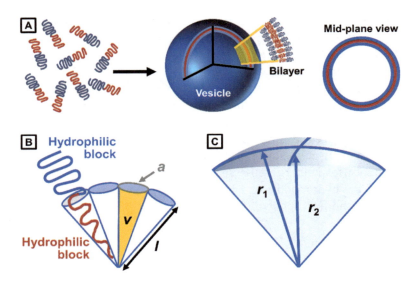

Figure 3.31: Vesicle formation by spontaneous self-assembly of amphiphiles and description of the shape of the self-assembled amphiphilic structures by the ratio of the amphiphile's hydrophobic-to-hydrophilic block length and its implication for the curvature of the hydrophilic–hydrophobic interface. (A) Schematic of amphiphiles (red: hydrophobic block, blue: hydrophilic block) self-organizing into a spherical vesicle. (B) Illustration of the amphiphile packing parameter P by the interfacial area, a, the amphiphile's hydrophobic volume, v, and its chain length normal to the interface, l. (C) Expression of the packing parameter by the radii r_1 and r_2. Adapted from [27]; copyright 2003 WILEY.

Exemplarily, in case of cylinders, $K = 0$, and $H = \frac{1}{2r} = \frac{1}{2l}$. Putting these into eq. 3.2 gives

$$P = 1 - \frac{l}{2l} + 0 = \frac{1}{2} \tag{3.3}$$

Due to the freedom of microfluidic device design, co-flowing multiphase flow is not limited to just two phases (e.g., hydrodynamic flow-focusing), but can also be extended to any higher number of fluid jets, as long as they can all be fitted into a microchannel cross-section. These multiple jets may be employed for separation purposes, for example, in microfluidic free flow electrophoresis. Multiple defined concentration profiles as found in such multiphase HFF are also suitable to locally regulate cellular functions, such as protein synthesis and morphogenesis. As just one example, consider the formation of multiple liquid jets for chemical induction and delivery of growth factors to surface-adherent cell lines, as shown in Figure 3.32A [28]. Another variation of CF microfluidics that is not based on HFF is found in co-flowing liquid interfaces in surface patterning for SF microfluidic applications, as shown in Figure 3.32B.

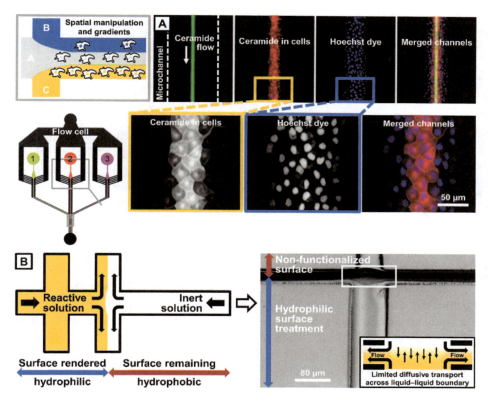

Figure 3.32: Spatially confined patterning of microchannel surfaces for localized cell culture treatment and surface modification of microchannels. (A) Spatiotemporal regulation of chemical gradients passed over cell cultures in conventional Petri dishes for localized induction and morphogenetic control by utilizing a microfluidic hydrodynamic flow-focusing device with multiple inflow channels (second row, left sketch). Fluorescent vital dyes Hoechst 33342 and ceramide are flow-focused into thin jets and passed over a kidney cell culture highlighting the spatial resolution of this method to be within a single-cell diameter. Adapted from [28]; copyright 2014 Royal Society of Chemistry. (B) Spatially localized microchannel surface functionalization to pattern a microfluidic device with two flow-focusing junctions for forming O/W/O double emulsions. Due to stable laminar flow during the surface treatment (bright-field microscopy image, right panel), the position of the liquid–liquid boundary of a reactive phase and a blocker phase only shifts within a micrometer range, thereby providing sufficient spatial resolution for controlling the channel surface wettability within a single cross-junction. By injecting a reactive solution via the designated inflow channel of the device, and an inert solution preventing the reactive solution from flowing into the designated outflow channel, the outflow channel remains hydrophobic. With this wettability pattern, oil-in-water (O/W) droplets may form in later SF experiments in the left part of the device, where the continuous aqueous phase wets the now hydrophilic microchannels before entering the right part of the device that remained hydrophobic during the surface treatment. Here, only a continuous oil phase can efficiently wet the microchannel surface, thus promoting the formation of oil-in-water-in-oil (O/W/O) double emulsions.

3.4 The practice of segmented-flow microfluidics

3.4.1 Device design and tailoring

The formation of single as well as multiple emulsions is one of the key applications of microfluidic flow cells in the broad variant of droplet-based or SF microfluidics. The richness of methods for fabricating these flow cells enables users to design and customize microchannels that provide precise control over the droplet size, the droplet shell thickness in the case of multiple emulsions, the droplet shape, and the droplet loading efficiency with additives. In particular, the inherently large surface-to-volume ratio of microchannels plays an important role in optimizing the contact area between fluids and channel surfaces, and thus the emulsion formation and stability in SF microfluidics. These interactions are usually tailored in terms of wetting and contact angle, as covered in greater detail below.

For SF formation, the choice of a pair of dispersed phase and continuous phase as well as a suitable surfactant that stabilizes the interface between these two phases if droplets are formed primarily depends on the solubility of to-be-used reagents (e.g., prepolymers, monomers, or analytes) and on their ability to be processed into droplets. In many applications, there may be further requirements or side conditions to be fulfilled, such as need for specific types of surfactants (e.g., biocompatible ones in biomicrofluidics) or use of specific solvent mixtures (e.g., in the case of block-copolymer handling). Beyond these general considerations, dynamic processes in microfluidic droplets and at their interface, such as those discussed in Section 1.5.4, strongly influence the emulsion dispersity and compartmentalization of solutes. These dynamics have important implications for the design of experiments in droplet microfluidics, as they may lead to transport of molecules into the continuous phase, and with that, to cross-contamination between the droplets, and thus to a loss of the defined experimental conditions associated with the discipline of droplet microfluidics. While microfluidic droplets are generally perceived as individual, independent reaction compartments, this picture only holds if experimenters choose the right continuous phase for a given dispersed phase and solute species, and also consider emulsion additives, for instance. In the former case, fluorinated oils such as Fluorinert™ FC-40 (3M™) and HFEs (3M™ Novec™, cf. Table 3.3) are frequently employed, because these oils are biocompatible, and because most biological and organic species are insoluble in these fluids, such that they provide a physical barrier for emulsion solutes. In the latter case, additives such as BSA or homopolymers based on one of the surfactant building blocks – for example, PLA added to PEG-*b*-PLA-stabilized droplets – coassemble at the liquid–liquid boundary to form a barrier-like viscoelastic layer that slows down exchange processes of solutes or improve the solubility of molecules in the dispersed phase.

Once a proper pair of fluids and further additives such as surfactants or downstream-addable reagents such as polymerization initiators or accelerators, all dissolved in suitable fluid media, have been identified, the greatest challenge in SF

Table 3.3: Typical fluorinated oils for droplet microfluidics and their chemical formula as well as boiling point (which is important in case of need for removal after the microfluidic experiment).

Fluorinated oil	Chemical formula	Boiling point (°C)
HFE™ Novec™ 7100	$C_4F_9OCH_3$	61
HFE™ Novec™ 7200	$C_4F_9OC_2H_5$	76
HFE™ Novec™ 7500	$C_7F_{15}OC_2H_5$	128
Fluorinert™ FC-40	$N(C_4F_9)_2CF_2CF_2H$	155
Fluorinert™ FC-70	$N(C_5F_{11})_3$	215
Fluorinert™ FC-72	C_6F_8	56

microfluidics is to devise a proper channel platform. The most common and most versatile variant is stamped flow cells consisting of fluid inlets and interfacing channel junctions, in which droplet formation takes place. As pointed out before in Chapter 2, a crucial parameter in this context is the channel surface **wettability**. This is because the most simple variant of stamped flow cell fabrication is soft lithography with a uniform channel height, thereby yielding microchannels in which a designated inner phase cannot simply be kept apart from channel-wall contact by geometrical features such as broad channel widening behind a channel junction. As a consequence, the channel-wall wettability has to be tailored such to allow good wetting by the designated outer phase in a channel junction, whereas the designated inner phase shall exhibit no wetting on the channel surface. If this is realized, then the outer phase will stay in contact with the channel walls, whereas the inner phase will tend to avoid such contact, thereby greatly facilitating its dispersion into droplets that are fully surrounded by the continuous phase. As a result, a channel junction that is supposed to form oil-in-water droplets must have hydrophilic walls, whereas one that is supposed to form water-in-oil droplets is must have hydrophobic walls. But what if both such junctions are present in a microchannel network? This is case if water-in-oil-in-water or oil-in-water-in-oil double-emulsion droplets are to be formed, where one droplet species is first formed in a first channel junction and then subject to a second dropmaking step in a second channel junction. In such devices, both hydrophobic and hydrophilic channel-wall wettabilities must be present in a spatially resolved patterning. The key to generating such spatially resolved wettability patterns relies on two steps. First, the native PDMS and glass channel walls are coated with a **sol–gel layer** that is then, second, subject to wettability modification by spatially confined **grafting polymerization**. Actually, sol–gel coatings in PDMS-based microfluidic devices serve even two main goals. First, these glass-like coatings are durable and resist harsh chemicals, temperatures, as well as mechanical load. Thus, sol–gel coatings are well-suited to provide mechanical stability to microchannels. For example, upon application of such a coating, the often problematic swelling of PDMS accompanied by a

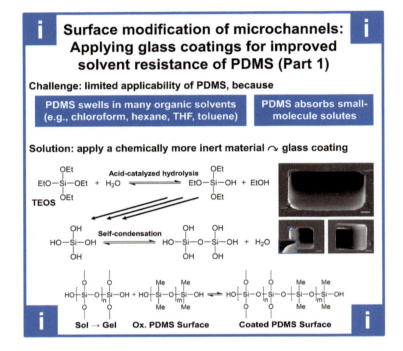

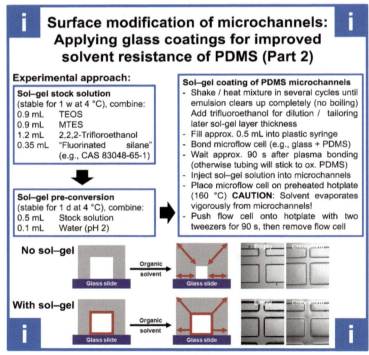

loss of structural stability that is observed in the presence of many organic solvents (e.g., hexane, chloroform, toluene) is prevented. Second, as already indicated above, sol–gel coatings may serve as a basis for spatially resolved channel-wall wettability patterning by grafting polymerization.

While sol–gel coatings can be purely made of tetraethyl orthosilicate (TEOS), in most recipes, it is common to employ a mixture of TEOS and triethoxy(methyl)silane (MTES) [29]. Depending on the molar ratio of TEOS and MTES, the sol–gel coating will be brittle and dense or flexible and porous. The reason is that TEOS contains four functional ethoxy groups that can participate in sol–gel network formation, whereas in MTES, a nonreactive methyl group replaces one ethoxy group. Thus, the more MTES the pre-sol–gel mixture contains, the less crosslinks form, such that the sol–gel becomes less rigid. This feature is crucial, as PDMS and glass-like sol–gels are conventionally quite dissimilative materials. If a native PDMS-microchannel is exposed to chloroform, it will get significantly narrower due to the swelling of the surrounding PDMS toward the channel center. A means to prevent that is to stabilize the PDMS channel walls by a sol–gel layer, but if that layer is too thin, it will delaminate from the PDMS surface and break instead of providing structural stability to the microchannel. Thus, more compliant sol–gel coatings, as achievable by a greater content of MTES in the pre-sol–gel mixture, are suitable to prevent deformation of the microchannels due to swelling or due to any other sort of mechanical stress exerted on them.[29]

In general, sol–gel coatings are neither perfectly hydrophobic nor hydrophilic, so in their native state, they only provide mechanical stability to the channels, but they do not contribute to optimized channel surface wettability. To also add that feature, an additional compound can be added to the pre-sol–gel mixture: either a silane compound with a perfluorinated carbon tail (usually 1H,1H,2H,2H-perfluorodecyltrimethoxysilane) or one with a hydrophilic polymer tail, for example, 2-[(acetoxy(polyethyleneoxy)propyl]triethoxysilane. With these additives, a sol–gel will be obtained that exhibits either highly hydrophobic or highly hydrophilic surface wettability, thereby enhancing the performance of the microchannels for W/O or O/W droplet formation.

---

29 The numbers provided for preparing sol–gel stock solutions as well as monomer / pre-polymer solutions towards functional microchannel coatings hereafter are more a rule of thumb than universally applicable. Depending on the total volume of the microchannel network that is to be coated as well as on the smallest feature size of the flow cell, the concentrations of TEOS and MTES as major components in the pre-sol–gel mixture need to be carefully adjusted, and with that its viscosity as well as the thickness and brittleness of the later sol–gel coating. To give a typical example, the numbers provided in the corresponding info boxes are optimized for fabricating planar PDMS-based flow cells with a spatially patterned sol–gel coating for forming double-emulsion droplets with an outer droplet diameter of approx. 100 μm.

The constraint of planar microflow cells as obtained by standard soft lithography, and its consequence on the need for tailored surface wettability in SF microfluidics becomes even more pronounced when microfluidic devices shall serve to produce for higher-order emulsions. These include double (W/O/W or O/W/O) and triple (W/O/W/O or O/W/O/W) emulsions, for instance. Here, each additional emulsion shell generally requires an additional droplet-forming microchannel junction with individual wettability; a hydrophobic one for forming an oil shell, and hydrophilic one for an aqueous shell. As a result, for instance, forming a W/O/W/O triple emulsion necessitates three flow-focusing junctions, the first being hydrophobic, the second being hydrophilic, and the third being hydrophobic again. This way, water droplets are encapsulated into an oil phase at the first junction, the as-formed W/O single emulsion is surrounded by an additional water shell at the second junction, and the as-formed W/O/W double emulsion is finally encapsulated in another oil phase yielding the desired triple emulsion, as shown in Figure 3.33. With regard to the typical dimensions of microchannel features in the range of 1 to 100 μm, their wettability has to be spatially localized with micrometer-scale resolution to ensure reproducible formation of emulsions with desired shell compositions. Due to the importance of single and higher-order emulsions for templating core–shell polymer particles or for serving as protective capsules for drugs, fragrances, and food, the preparation of corresponding spatially patterned flow cells will be in the focus of the following. Here, we will look at three patterning methods: surface modification of planar[30] microchannels with hydrophilic polymers via grafting-from and grafting-to process employing localized UV light or inert blocker phases inside the microchannels, sequential layer-by-layer deposition of polyelectrolytes, and localized surface oxidation via plasma discharge [30–33].

We start with the simplest method of surface patterning, which utilizes **plasma oxidation** for controlling the microchannel wettability. In Section 3.1.6, we have

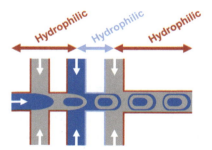

Figure 3.33: Schematic of a microflow cell for preparing W/O/W/O triple emulsions. The required surface wettability at each flow-focusing junction is highlighted in light blue (hydrophilic) or red (hydrophobic).

---

[30] Of course, all processes towards tailored surface properties with spatial control described in this chapter are also suited for microchannel modifications in multi-layered flow cells.

discussed the technique of plasma oxidation for the purpose of PDMS-to-substrate bonding in the process of microfluidic device fabrication already. During such treatment, silanol groups are generated on the PDMS and also on glass if that is used as the device substrate (which it commonly is). These groups are hydrophilic; hence, the plasma treatment inherently renders the channel-wall surface hydrophilic. The only issue is that this hydrophilic functionality will degrade over time due to follow-up reactions with air, but in a short-term view, the plasma oxidation is a simple and quick method to generate channel surface hydrophilicity. Hence, if a whole device shall be rendered hydrophilic, and if it is fine to do so only on a short-term perspective, simply subjecting the whole device to plasma is a quick and easy method. Such treatment can also be performed in a somehow spatially resolved fashion by blocking the inlets of the channels that shall not be treated, thereby preventing the plasma to enter these channels [33]. A different method with similar outcome is layer-by-layer assembly of polyelectrolytes onto silanol-activated PDMS channel surfaces, rendering them strongly hydrophilic [32]. This wettability modification is more durable than simple plasma activation, although it may also vanish over time due to delamination.

As simple as the two preceding methods are, they are not quite versatile, as they allow channels to be rendered hydrophilic only, but not hydrophobic; they also provide no or at most just poor spatial resolution. A more sophisticated method that overcomes these limitations is based on **sol–gel coatings** with further functionalization. There are two variants to achieve that: grafting-to polymerization and grafting-from polymerization onto the sol–gel layer. Both rely on an initial sol–gel coating that is either native or rendered hydrophilic already by inclusion of a silane compound with a perfluorinated tail (usually 1H,1H,2H,2H-perfluorodecyltrimethoxysilane). In addition, at least one further component is added to the sol–gel, which is a silane functionalized with a reactive carbon–carbon double bond, for example, 3-(trimethoxysilyl)propyl methacrylate; some researchers also include a silane-modified photoinitiator into the sol–gel to activate this bond by light [30], but this can also be achieved by filling the channels with a photoinitiator solution, or alternatively, with a solution of a thermally labile initiator. The double bonds on the surface of the sol–gel layer may serve as an anchor to graft-to or graft-from hydrophilic polymer chains such as polyacrylic acid, which then serve to attach a "hydrophilic lawn" onto the sol–gel surface. This "lawn" renders the surface wettability to be strongly hydrophilic. If this is done allover the sol–gel coating, the whole device will be hydrophilic, whereas if this surface modification is done in a spatially resolved fashion, only localized hydrophilic patches can be generated to provide spatially resolved hydrophilic wettability. This method is of prime use for generating high-resolution wettability patterns in the microchannel network, as it is needed for high-quality multiple-emulsion formation. The spatial confinement in the grafting-to or grafting-from process can be achieved by two techniques: **photopatterning** [30] or **flow confinement** [31], as detailed in the info boxes below.

Surface modification of microchannels: Photo-reactive and other functional coatings for spatially patterned wettability

Beyond durability and resistance against organic solvents, sol–gel coatings may also serve as substrates for **precisely tailored microchannel wettability** by graft-polymerization. For that, conventional sol–gel coatings are further functionalized with photo-reactive silanes (**Method 1 and 2**).
If the intrinsically low solvent resistance of PDMS or pronounced diffusion of small solutes into PDMS do not need to be taken into account, polymer coatings for spatially patterned microchannel wettability may also be directly applied onto plasma-activated PDMS surfaces (**Method 3**).

Three examples:

METHOD 1: Grafting-to functionalization	METHOD 2: Grafting-from functionalization	METHOD 3: Layer-by-layer assembly
PDMS with sol–gel	PDMS with sol–gel	Plasma-treated PDMS

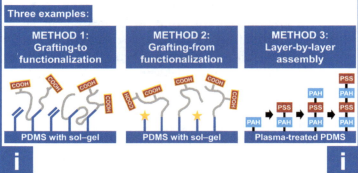

Surface modification of microchannels: Locally restricted surface reactions by spatially patterned UV light and flow confinement

For higher-order emulsion formation in planar PDMS-based microflow cells (e.g., O/W/O droplets), it is crucial to locally modify the interfacial chemistry of the microchannels with high contrast. Depending on the quality of the optical setup and the stability of the liquid–liquid interface of the reactive / inert phase, respectively, the resolution of the graft polymerization can be as fine as 1 μm.

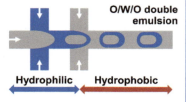

O/W/O double emulsion

Hydrophilic — Hydrophobic

Flow confinement — **Spatial UV illumination**

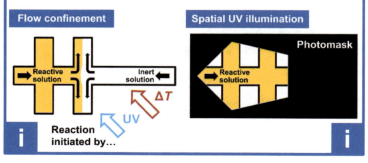

Reaction initiated by...

The flow confinement technique relies on flowing a solution of a hydrophilic compound that can be attached-to or grown-from the double bonds on the sol–gel layer surface through the channels that are supposed to be rendered hydrophilic. Most typically, the reactive compound in that solution is acrylic acid along with thermally or photochemically labile initiators (if such are not already embedded in the sol–gel layer as well). The channels that are not supposed to be rendered hydrophilic, but instead, supposed to stay hydrophobic as given by the hydrophobically modified sol–gel layer, is flown-through by an inert blocker solution, most typically water. Thermal or photochemical initiation of the process of grafting hydrophilic polymer chains to or from the double bonds on the sol–gel layer in those regions where the reactive solution flows then serves to create a spatially resolved wettability patterning. Most typically, this is realized by grafting chains of polyacrylic acid from the double bonds through photochemically or thermally induced free-radical polymerization of acrylic acid monomers in the reactive solution. During that process, the inert blocker phase and the reactive solution form a stable interface at locations within the microchannel network that depend on how the two fluids are injected into the to-be modified flow cell, as shown in Figure 3.34A and B and detailed further in the info boxes below. A challenge in this approach is that there can easily be slight fluctuations of the flow front, such that micrometer-precision at the cross-junction is challenging to achieve, as seen in Figure 3.34C.

The flow rates of the reactive and the blocker fluids determine the interface between them in the region of contact, and they also determine the average residence time of the reactive molecules in the microchannels.[31] Often, manual readjustment of the flow rates is necessary during the process, for example, due to the increase of viscosity of the reactive solution due to in-bulk polymerization of its reactive monomers in addition to the desired on-surface polymerization on the sol–gel-coated microchannels.

Initiation of the grafting reaction can either be done by photochemical or by thermal triggering. For the thermally driven reaction, we require a hot plate to heat the to-be-processed microfluidic chip once a desired flow-confinement flow pattern has been established in it, along with an upright microscope on top to observe the flow front. For the UV-initiated reaction, we require an optical setup, for example, a flood UV system or black light with the correct wavelengths. A drawback to this approach is that, without a microscope, it will not be possible to see the progressing microchannel surface polymerization. Another drawback is that such systems unlikely provide high light intensity. To overcome both limitations, we recommend to use a commercially available fluorescence microscope, which is equipped with a notably

---

[31] The values given in the following info boxes are based on the authors' experience; they may serve as some initial guidance, but need to be adjusted in every actual experiment, as they are highly dependent on the microchannel cross-section as well as on the total volume of the flow cell, and they also need to be adjusted regarding the change of viscosities and concentrations of the fluids *during* the grafting process.

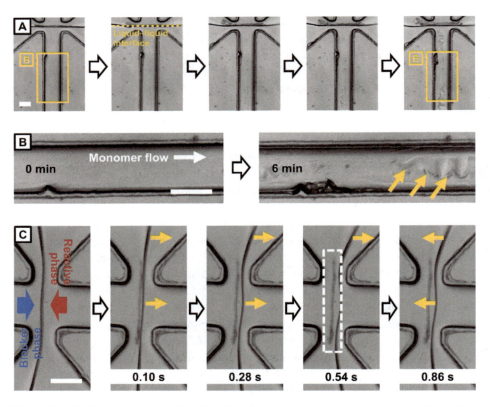

Figure 3.34: Spatially controlled wettability patterning of sol–gel-coated microchannels utilizing flow confinement of reactive phases. (A) Bright-field microscopy image of a microchannel cross-junction. A liquid–liquid interface forms in the center of the cross-junction (highlighted by an orange dotted line) if water is acting as an inert block phase and an aqueous monomer or prepolymer solution is acting as reactive phase, injected from the top and the bottom of the junction, respectively. In the lower part of the microfluidic device, which is flown-through by the reactive liquid phase, a layer of hydrophilic polymers forms on the sol–gel-coated microchannel surface over time. (B) Zoom onto the sol–gel-coated microchannel, which is functionalized with a hydrophilic polymer coating by the grafting-to method. As the coating is sufficiently dense, typical wrinkling is observed (highlighted by arrows), which can be utilized as an end-point indicator for sufficient coating thickness in the examined area. (C) Liquid–liquid interface dynamics within the microchannel cross-junction. Flow-rate fluctuations as well as the increase of the viscosity of the reactive monomer or prepolymer phase during the process, which is caused by in-bulk polymerization of the reactive monomers parallel to their desired on-surface polymerization, causes the liquid–liquid interface to move back and forth (orange arrows) within the junction during the procedure. The scale bars denote 50 μm.

strong UV-plus-visible light source by default. With such a microscope, the strong and UV-containing illumination light can be focused onto a narrow spot, which is then present at very high local intensity, and the fluid flow front in that region can be visualized in parallel, thereby allowing it to be monitored during the grafting reaction and to account for potential drift by manual flow-rate adjustment. An example for such an experimental basis is an Olympus IX81 microscope equipped with a 100 W (or more) mercury lamp (365 nm, e.g., USH-1030L, USHIO Inc.). If photoinitiators with significantly lower excitation wavelength are considered, it is advisable to exchange all lenses along the optical path of the illumination setup by UV-transparent fused silica, which, however, naturally requires higher investments in the experimental setup. As an alternative to an expensive fluorescence microscope, users may also build their own setup composed of a standard optical microscope (e.g., a Zeiss Axiovert) that uses just a common light bulb or weak diode for the illumination and add to that a homemade optical path for strong UV-light in-coupling. Such a setup is shown in Figure 3.35.

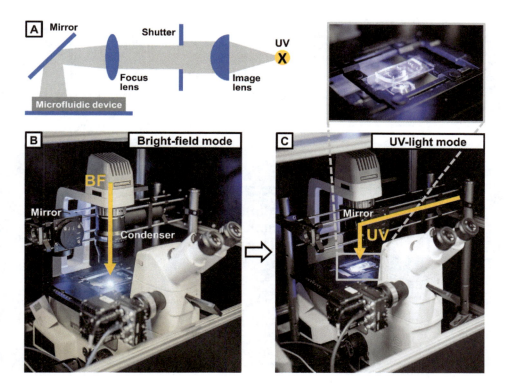

Figure 3.35: Self-built experimental setup for spatial patterning of microchannel wettabilities via UV-triggered grafting on their sol–gel-functionalized surfaces. The setup is based on a conventional inverted microscope to which a UV source (e.g., OmniCure® S1500, 200 W nominal output, approx. 10 W cm^{-2} maximum surface power density) is coupled. (A) Sketch of the optical path for focusing a microscopic UV spot onto the microchannels of a flow cell mounted onto the microscope stage. By solely utilizing fused silica lenses, a variety of photoinitiators with excitation wavelengths from hard UV light to the visible range (approx. 280 to 520 nm) may be employed. (B) Setup operated in bright-field mode for aligning the flow cell prior to UV illumination. (C) In the UV-light mode, the condenser that is part of the bright-field optical path above the microscope table is removed; instead, a mirror is moved over the microscope table to allow for exposing the flow cell (enlarged view above (C)) by a focused UV-light beam coupled into the experiment from the side.

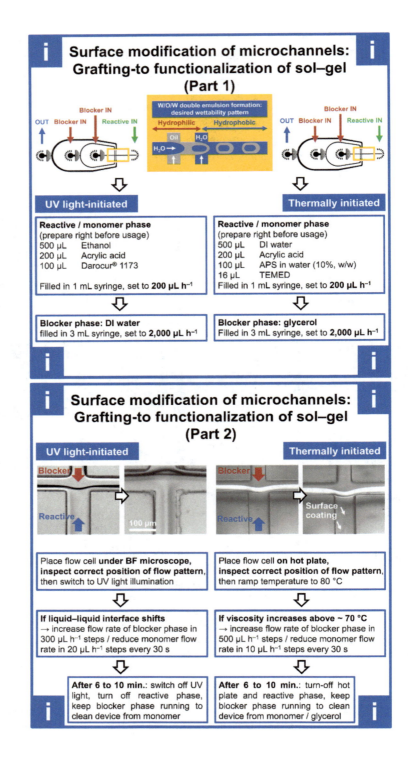

A setup like the one shown in Figure 3.35 provides general utility for photochemical modification of microchannel surface chemistries, not only in the variant of flow confinement, but also in very general, because the possibility to strongly focus a spot of UV light onto specific areas of a microchannel network can generally serve to locally modify the surface chemistry exclusively there. If the light focusing is very sharp, then we may even leave away the auxiliary condition of flow confinement, but instead, simply park a reactive solution in the microchannels and let it react in a locally confined fashion by sharply focused light triggering. To make the microchannel surface modification even better resolved, use of an immobilized photoinitiator prohibits smearing of the photo-pattern by diffusive motion of activated species into non-illuminated regions. Such immobilization can be achieved by use of a silane-functionalized photoinitiator that can be made part of the sol–gel coating on the microchannels [30].

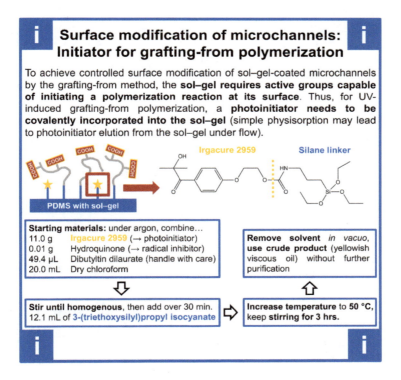

So far, we have covered two methods to tailor the microchannel surface chemistry with micrometer precision, particularly the wettability. In both, we may either use a photoinitiator incorporated into the sol–gel layer, thereby enhancing the spatial resolution of the photopatterning process, or we may use a photoinitiator dissolved in the reactive monomer or precursor polymer solution. The latter approach has two practical advantages. First, unlike surface-immobilized photoinitiators, there is

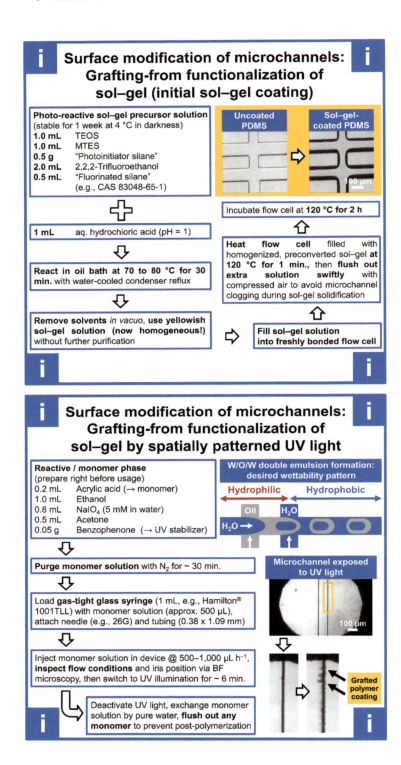

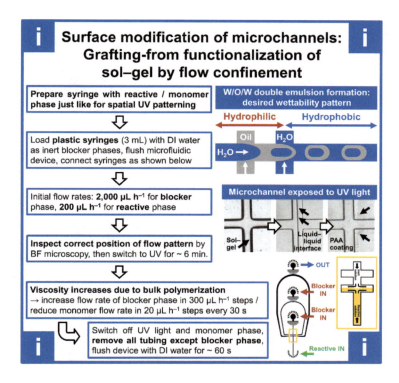

a huge number of photoinitiators available for bulk solutions, which can be constantly refreshed by the flow if the method of flow confinement is applied, thereby constantly refreshing the photoinitiator within the UV-exposed areas. Thus, the reaction time, and with that the amount of polymer grafted to the surface, can be controlled by the flow velocity and the duration of flow. Second, the solution-based approach also benefits from a large structural variety of photoinitiators soluble in reactive monomer solutions. By contrast, the approach of using sol–gel-immobilized initiators requires a structural motive on the initiator that can be coupled to the sol–gel prior to the grafting process, usually a silane group, which is hardly found in commercial photoinitiators and often needs to be introduced via Schlenk synthesis technique (cf. info box entitled Initiator for grafting-from polymerization).

For users new to the field of microfluidics, patterning a microfluidic device via grafting-to or grafting-from polymerization is challenging and requires extended experimental equipment (e.g., a UV-light setup or an upright microscope) that is not available in every research group potentially interested in complex microfluidic devices. To circumvent these challenges, an alternative method may serve useful: **layer-by-layer (LbL) assembly** of charged polymers to render a (previously activated) channel surface hydrophilic. The LbL method is based on chemi- or physisorption of a stack of alternatingly charged polyelectrolyte layers, which are deposited one after and on

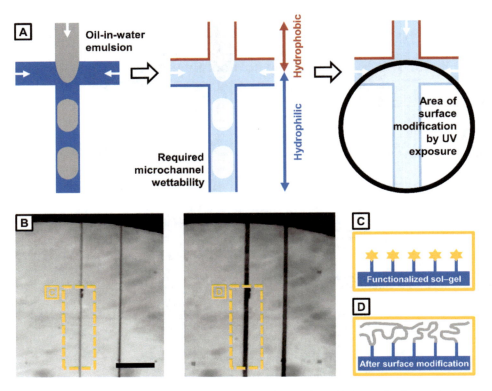

Figure 3.36: Surface modification of PDMS-based microfluidic channels by covalently attached polymer coatings. (A) From left to right: Microfluidic experiment for forming oil-in-water droplets utilizing a planar microfluidic cross-junction. For that purpose, we require a local switch of the wettability from hydrophobic to hydrophilic at the cross-junction. Only then, oil droplets will be surrounded by the aqueous continuous phase wetting the surface of the outflow channel walls. To locally control the microchannel wettability, one approach involves the spatially controlled exposure of the microchannels, which are coated with a functional glass-like sol–gel coating allowing for grafting-to or grafting-from polymerization of hydrophilic polymers. For that, a CAD-based photomask or optical iris is placed between a high-power UV source (e.g., 100 W–200 W nominal output) and the microfluidic device, thereby restricting the area of UV illumination and surface modification, respectively. (B) Bright-field microscopy images of a sol–gel-coated microfluidic channel with the edge of an iris visible in the upper part. As the functionalized outflow channel surface is exposed to UV light in the presence of polymerizable hydrophilic monomers or macromers continuously flowing through the microchannel (C), the wall thickness increases due to an as-formed hydrophilic overlay (D). The scale bar is 100 µm.

top of another. This technique is simple and can be applied even to non-sol–gel-coated but simply plasma-activated PDMS surfaces, where it provides strong hydrophilic wettability. In contrast to the sol–gel-based techniques of grafting, however, it does not provide strength of the channels against organic solvents, and its ability to provide spatially controlled wettability adjustment is not as versatile as that of the grafting techniques on sol–gel.

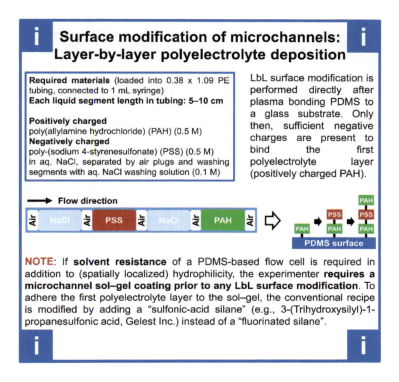

After completion of the wettability-treatment procedure, we need to check its outcome. This is crucial to assess the correct position of the wettability pattern and to check for sufficient coverage of the microchannels with a functional surface coating. We have two methods at hand to perform such checking; both are based on optical inspection. First, to check for the performance of the sol–gel covering and the polymer-grafting on top of it, we can apply both to just a simple glass slide and then let a water droplet wet it. Whereas the droplet should poorly wet the intrinsically hydrophobic sol–gel-coated glass surface, it should strongly wet the same surface after post-functionalization with hydrophilic poly(acrylic acid) (PAA), as shown in Figure 3.37A. The latter surface may be further characterized by subjecting it to the acid-sensitive dye toluidine blue, which stains only the PAA-covered parts, as shown in Figure 3.37B. This way of analysis may also be applied to microscopic channels in a microflow cell, in which the toluidine blue dye stains the hydrophilic PAA-grafted parts only, as seen in Figure 3.37C, left. Alternatively, the location of the wettability crossover within a microchannel may be determined by the formation of a meniscus of water flushed into the spatially patterned microchannel (Figure 3.37C, right).

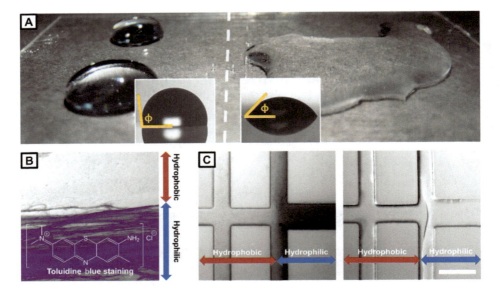

Figure 3.37: Optical inspection of modified surface wettabilities in microfluidic flow cell components. (A) Wetting of a water droplet on an intrinsically hydrophobic sol–gel-coated glass surface (left) and on the same surface after post-functionalization with hydrophilic poly(acrylic acid) (PAA) (right). The insets show the corresponding contact angle measurements by the droplet shape method. (B) Glass surface partially covered with PAA, visualized by staining with the acid-sensitive dye toluidine blue. (C) Characterization of locally patterned surface wettabilities in a microfluidic devices under static conditions (no flow) by staining the hydrophilic PAA-grafted regions with toluidine blue (left) and by determination of the meniscus at the wettability crossover. The scale bars are 200 μm.

3.4.2 Formation of W/O microdroplets and polymer microparticles

With a tailored microfluidic device at hand, we can go ahead and form microscopic droplets with tailored size and practically monodisperse distribution. These droplets may either be used as they are, for example, acting as femtoliter-scale reaction vessels for downstream (bio-)analytics, or they may serve as templates for further preparative steps, most prominently droplet curing or gelation to obtain (polymer-based) microparticles or microgels. The first step in all that is to form a monodisperse emulsion. For this purpose, we need a microfluidic chip with a single channel junction, either stamped into a PDMS elastomer in the form of a four-way cross, as detailed in Sections 3.1.2–3.1.6, or assembled from two coaxially aligned glass microcapillaries, as detailed in Section 3.1.7. Pumping two immiscible fluids into this device leads to droplet formation at the junction, which will occur in the form of controlled dripping at adequate flow conditions, precisely, at Weber and Capillary numbers both not too far above one (cf. Section 1.5.3). In glass-capillary microfluidics, such an emulsion can basically be formed from any pair of nonpolar oil and water, ranging from volatile organic solvents such as chloroform or hexane to viscous paraffin oils. In PDMS-based stamped flow cells, by contrast, volatile organic fluids are unsuitable, as they tend to swell PDMS and thereby lead to channel-narrowing or even clogging. As a result, nonvolatile and more viscous paraffin oils are more suitable there. The most ideal fluid for emulsion formation in general, both in glass-capillary microfluidics and especially in PDMS-based microfluidics, is fluorinated oil such as 3M's HFE 7500. The reason for this explicit recommendation is that this oil is highly immiscible with water, thereby leading to well-defined oil–water interfaces, it has well-defined Newtonian flow characteristics, it is nontoxic, and it has a high permeability for gases, such that precious objects like living cells inside the droplets (which may be encapsulated in these because they are supposed to be subject to downstream bioanalytics or to further following microgel entrapment) may breathe through the surrounding continuous oil phase. On top of that, the refractive index of fluorinated oil is very different from that of water, such that the resulting emulsion can be clearly visualized in a microscope. The only drawback of fluorinated oil is its rather high price, and, in some cases, its low viscosity; even though the latter property is often, on the contrary, very beneficial, there may be situations where it is not, for example, if an aqueous polymer solution is to be emulsified, which has to occur by an outer fluid with high viscosity and thereby high enough external shear.

With a microfluidic channel junction and two suitable immiscible fluids, we may form droplets with highly controlled sizes and monodisperse distribution if we operate our device in the controlled dripping regime, as detailed in Section 1.5.3. Without further additives, however, these droplets will soon undergo coalescence to minimize their interfacial area, and with that, the interfacial tension, as shown in Figure 3.38A. To prevent that, the droplets' water–oil interface must be stabilized by a **surfactant**. The easiest way to achieve such stabilization is to add a suitable surfactant to the external, continuous phase (for example, the oil phase in the case of W/O droplet formation), from which the amphiphilic surfactant molecules may adsorb at the water–oil interface and thereby stabilize the droplets. At that condition, no droplet coalescence will occur downstream, not even if the droplets come into direct contact, as they do in channel regions where their initial single-file and mutually separated flow changes to one where they get contact to each other, as seen in Figure 3.38B.

Note: the approach of adding a surfactant along with the continuous oil phase (which means, premixed to it prior to the microfluidic experiment) is convenient. A little drawback of that approach, though, is that the therewith inherently reduced interfacial tension will not only lead to stabilized droplets (as desired), but also to higher *Ca* and *We* numbers, as the interfacial tension enters both these quantities in their denominators. As a result, transition from a controlled dripping state to a less controlled jetting regime will occur at lower flow rates than it would if the interfacial tension was higher. Hence, when an experiment is targeted at

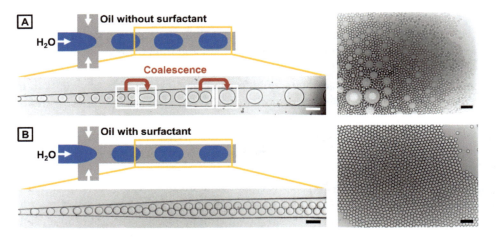

Figure 3.38: Influence of surfactant on microfluidic droplet formation and stability for the example of water-in-oil emulsion formation. (A) In a typical cross-junction flow cell, microfluidic generation of a monodisperse emulsion (here: W/O) creates droplets that are initially separated by oil plugs. These single-file flowing droplets, however, usually close up and align in a pearl necklace-like assembly downstream, when microchannel widens (which it will latest do in the exit tubing region of the device). Without chemical stabilization, the droplets undergo coalescence in this situation as well as after collection (right image). (B) By contrast, emulsion droplets that are covered by a surfactant, which is commonly dissolved in the oil phase, are stabilized against coalescence, both in widening channels on the chip and after collection off the chip (right image). The scale bars in (A) and (B) are 50 µm in the left images, and 100 µm in the right images.

high productivity and throughput, it is desirable to form the droplets at a high interfacial tension such to allow high flow rates to be used while still keeping Ca and We below one, followed by immediate interfacial-tension lowering for droplet stabilization downstream. This can be realized in a microfluidic device with two sequential cross-junctions, one forming the droplets without surfactant in the oil phase and the other adding oil loaded with surfactant to stabilize the just-formed droplets [34].

Suitable surfactants for fluorinates oils contain at least one fluorinated side group or tail to allow for solubilization. Users may choose from a large library of fluorinated surfactants with diverse functional groups [35, 36]. Among these, the most frequently used fluorinated surfactant in (bio-)materials research are triblock-copolymer surfactants of type ABA, where A is a perfluoropolyether (PFPE, DuPont®, or Krytox® FSH/FSL) and B is poly(ethylene glycol) (PEG). One key point in the surfactant selection is that even though fluorinated surfactants may seem to be rather inert due to their selective solubility in inert fluorinated oils, though, nonspecific protein adsorption has been frequently observed [37]. If issues like that are not of relevance in a simple microfluidic experiment (which actually means, in an experiment that does not involve proteins or other sensitive ingredients), users may even circumvent the rather high price for specifically tailored commercial fluorinated surfactants by simply making their own. In an info box on that matter, we provide a simple recipe.

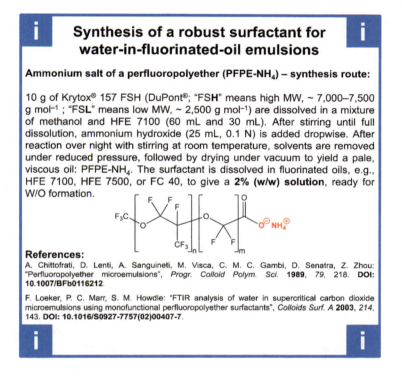

i **Synthesis of a robust surfactant for water-in-fluorinated-oil emulsions** **i**

Ammonium salt of a perfluoropolyether (PFPE-NH$_4$) – synthesis route:

10 g of Krytox® 157 FSH (DuPont®; "FSH" means high MW, ~ 7,000–7,500 g mol^{-1} ; "FSL" means low MW, ~ 2,500 g mol^{-1}) are dissolved in a mixture of methanol and HFE 7100 (60 mL and 30 mL). After stirring until full dissolution, ammonium hydroxide (25 mL, 0.1 N) is added dropwise. After reaction over night with stirring at room temperature, solvents are removed under reduced pressure, followed by drying under vacuum to yield a pale, viscous oil: PFPE-NH$_4$. The surfactant is dissolved in fluorinated oils, e.g., HFE 7100, HFE 7500, or FC 40, to give a **2% (w/w) solution**, ready for W/O formation.

References:
A. Chittofrati, D. Lenti, A. Sanguineti, M. Visca, C. M. C. Gambi, D. Senatra, Z. Zhou: "Perfluoropolyether microemulsions", *Progr. Colloid Polym. Sci.* **1989**, 79, 218. DOI: 10.1007/BFb0116212.

F. Loeker, P. C. Marr, S. M. Howdle: "FTIR analysis of water in supercritical carbon dioxide microemulsions using monofunctional perfluoropolyether surfactants", *Colloids Surf. A* **2003**, 214, 143. DOI: 10.1016/S0927-7757(02)00407-7.

With a suitable microfluidic device and fluid pair at hand (including a surfactant), we can generate microfluidic droplets by operating our device in the controlled dripping regime. This regime is present at *Ca* and *We* numbers both below one. In standard microchannels with diameters in the range of 20–200 μm, when we use fluorinated oil (plus about 1–2% (w/w) of surfactant) as the outer phase and water as the inner phase, this corresponds to flow rates in the range of 1,000–10,000 μL h^{-1}. The ratio of the inner:outer fluid volumetric flow determines the droplet size. In general, the higher the inner:outer fluid flow-rate ratio (FRR) is, the larger will be the droplets, and vice versa, as seen in Figure 3.39. A very common ratio in standard droplet-microfluidic experiments is 1:2 or 1:3. At a fixed FRR, the total flow rate of the fluids determines how many droplets are produced per time. Depending on the droplet size, this droplet production frequency can span from a few Hertz up to the kilohertz regime in usual microfluidic experiments. With such high productivity, we easily produce thousands to billions of droplets in an experiment. A crucial issue then is to adequately analyze these large ensembles after their formation such to quantify the average droplet size and its distribution. This is best achieved by taking adequate fluorescence or bright-field micrographs of several aliquots of the resulting emulsion, each recorded several times at different positions, and then to analyze them by software algorithms such as ImageJ plugins for this purpose. In an info box, we provide some details on how a good image should look like for this purpose.

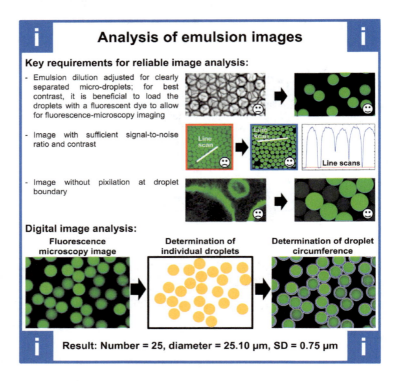

A further method for adjusting the droplet size and number in a microfluidic experiment is droplet splitting after the droplet formation, as schematized earlier in an info box in Section 2.5 and shown for an actual example in Figure 3.40 [38]. In this method, an initial large parent droplet size is split to smaller daughter-droplet sizes by letting the channel-filling droplet hit Y-junctions, where the droplet volume reduces to one-half in each junction. With that, the droplet size reduces by $(1/2)^{1/3} \approx 0.8$ in each such event, whereas the droplet number doubles.

Note: when imagining a microfluidic experiment or sketching it at a drawing board, microfluidic emulsion droplets are usually perceived to be perfect spheres with a diameter equal to the microchannel height and width (in case of squared microchannels) at the droplet-forming nozzle. However, as shown in Figure 3.39, when investigating an actual experiment via bright-field or fluorescence microscopy, the droplets formed in rectangular shaped PDMS-stamped microchannels

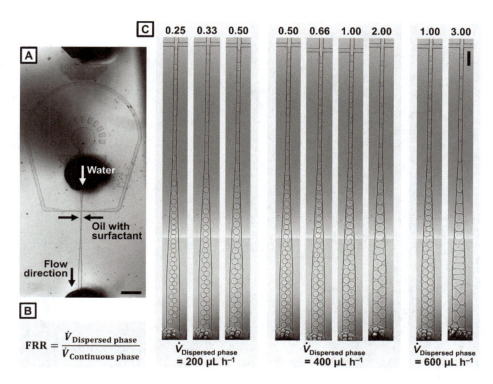

Figure 3.39: Formation of water-in-oil droplets, and control over the droplet size by the flow-rate ratio (FRR) of the inner:dispersed phase and the outer:continuous phase. (A) Bright-field microscopy image of a PDMS-based microfluidic device with a flow-focusing junction for droplet formation. The dark areas correspond to the punch holes and fluid inlet ports, respectively, through which the aqueous and oil phase are injected into the microchannels and through which the resulting emulsion is ejected. The scale bar is 500 μm. (B) Equation for calculating the FRR. (C) Droplet size depending on the FRR. The scale bar denotes 100 μm.

usually exhibit a more tic-tac- or sausage-like shape within the confinement of the channel. For yielding perfectly spherical droplets, it is therefore suggested to gradually increase the width of the microchannel until the as-formed droplets can relax into a spherical shape in thermodynamic equilibrium, which then, however, may turn out to be slightly larger than the droplet-forming nozzle (cf. Figure 3.39). In glass-capillary microfluidics, by contrast, the droplets usually have spherical shapes right from the point of their formation on, because the glass-capillary-based channels are usually much wider than the droplet diameter.

The droplet appearance in a microfluidic channel has important implications for the droplet utilization in potential downstream on-chip applications. In the following sections, we discuss a key area of such an application: the templating of microparticles from droplets. In this application, the droplet shape is fixed by

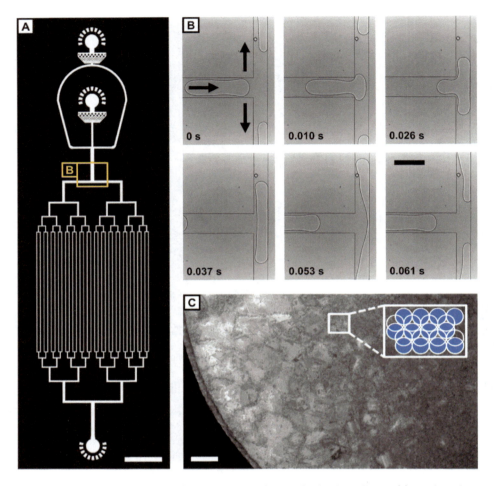

Figure 3.40: Downstream microfluidic emulsion modification by droplet splitting. (A) CAD-based photomask of a droplet-splitting device. The scale bar is 5 mm. (B) Image sequence of a water droplet splitting at a T-shaped channel into two halves. The scale bar denotes 400 μm. (C) Emulsion product after five splitting steps. The monodisperse W/O emulsion product assembles into a hexagonal packing. The scale bar is 1 mm.

downstream solidification or gelation of the droplets. As an inherent consequence of that solidification, the droplet shape directly translates into the follow-up microparticle shape. Hence, in this case, microfluidics users need to carefully pre-anticipate their desired droplet sizes, and with that, the one of the droplet-templated follow-up materials, in the actual digital layout of the microfluidic device beforehand.

As a simple example, we consider the case of droplet gelation to form micrometer-sized polymer gels, which are referred to as microscopic hydrogels or

microgels.[32] Most typically, these microgels are hydrogels, as obtained by gelation of W/O droplets by forming a percolated polymer network inside each droplet. This can be achieved if the droplets contain a prepolymerized polymer that can be crosslinked to obtain a three-dimensional network, or even simpler, if the droplets contain monomers that can be both polymerized and crosslinked in a single step. A good system to start with for practicing is to use the monomer acrylamide (AAm) and the crosslinker N,N'-methylenebisacrylamide (BIS), typically in a molar ratio of 99:1 and at a total concentration of 100 g L^{-1}. To initiate the polymerization, a further compound in the droplets must be a radical initiator such as ammonium persulfate (APS), in a typical concentration such as 0.5 mol% relative to the total monomer and crosslinker content. To accelerate the decomposition of that initiator, and therewith the crosslinking copolymerization of the monomers, a catalyst, N,N,N',N'-tetramethylethylenediamine (BIS), is added to the continuous phase, typically in a content of 1–5 wt.%. This compound is soluble in both oil (both hydrocarbon and fluorinated oil) and in water, such that it will enter the aqueous droplets from the continuous oil phase by diffusion and therein catalyze the decomposition of the initiator, which then triggers the polymerization at room temperature. This process turns the water droplets into microscopic polyacrylamide hydrogel particles. After letting the collected emulsion stand for some hours to ensure completion of the gelation process,[33] the product microgels can be isolated from the continuous phase and transferred into plain aqueous medium by a sequential washing procedure, as detailed in one of the following info boxes. Figure 3.41 shows a typical outcome.

Thermal initiation of droplet solidification, as introduced above, provides a simple path toward defined polymer hydrogel particles. Here, all components (initiator, prepolymer or monomer, and crosslinker) can be premixed and stored for the time of the microfluidic experiment at room temperature in a single fluid reservoir for the dispersed phase, thus requiring just one additional fluid reservoir for the continuous oil phase. For new microfluidics users, such systems provide a smooth access to material synthesis by microfluidic emulsions, before moving to more delicate polymerization reactions (e.g., biorthogonal Diels–Alder [4+2] cycloaddition, thiol-Michael addition, or azide–alkyne click reaction), where the reaction partners must be stored at fridge temperature and be kept separated from each other as long as possible before becoming mixed in the droplets.

So far, we have discussed how spherical droplets translate into spherical microparticles. As a next step, let us think about the use of nonspherical droplets for preparing polymer particles with more complex shapes, resembling those found in the amazing structural variety in nature. This can be achieved by adjusting

---

[32] For a definition of the term microgel, see Chapter 2.4.2.
[33] During that, the collected emulsion should be covered with light mineral oil if fluorinated oil is used as the continuous phase; this prevents drying of the gelling droplets, which float on top of the fluorinated oil phase due to their lower density.

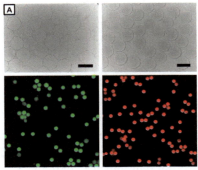

Figure 3.41: Microfluidically prepared, fluorescently labeled microgels made from poly(acrylamide) (PAAm). (A) Bright-field and fluorescence micrographs; the scale bars in the upper images are 25 µm. (B) Two batches of densely packed PAAm particle suspensions labeled with two different concentrations of Alexa 488 and Alexa 637, respectively.

the inner-phase flow rate such to yield droplets with elongated shape inside the confinement of the outflow channel behind a droplet-forming nozzle. This one-dimensional droplet squeezing yields nonspherical, elongated entities, and solidification of this out-of-equilibrium shape will fix it. Even more, preparation of flat, disk-like droplets is possible by simply squeezing them in not only one but in two dimensions in the outflow channel and then solidifying them. The only necessity in both these latter two approaches is to have a method that very rapidly solidifies the droplets in a defined area of the microfluidic chip, namely, in the confining microchannel right downstream of the droplet-forming channel junction. A method that provides such rapid droplet solidification is fast UV-initiated polymerization. For this purpose, we may use the same experimental setup that has been previously introduced in Section 3.4.1 for microchannel surface patterning. In such a setup, focusing of a strong UV beam onto the confining outflow microchannel will cause huge UV intensities there, leading to very rapid polymerization of suitable monomers inside the droplets. The same platform may also serve to form anisotropic microparticles, such as so-called **Janus particles**. These particles consist of two distinguishable halves, like the two-faced Roman god Janus. They can be made by first co-flowing two miscible (like two water-based) polymerizable or gellable fluids and disperse this co-flowing stream to droplets by flow-focusing with a continuous (like an oil) phase, as sketched in Figure 1.23C. The resulting droplets will retain the two-halved arrangement of fluids, and rapid solidification of the droplets will arrest that pattern such to obtain two-halved particles, named Janus particles.

Microfluidic experiment: Photopolymerization of microfluidic emulsions (Part 1)

A key application of droplet microfluidics is the preparation of microscopic polymer particles with defined size and internal organization.

By employing **UV light as external polymerization trigger**, a segmented flow of pre-polymers or monomers can be solidified **on-chip** right after droplet formation. That way, non-equilibrium structures can be generated and fixed by rapid solidification of deformed pre-particle droplets in a confining channel section, whereas conventional off-chip solidification of undeformed droplets yields spherical particles.

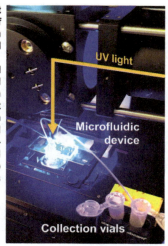

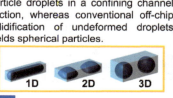

Microfluidic experiment: Photopolymerization of microfluidic emulsions (Part 2)

Example: Preparation of pancake-shaped poly(acrylamide) (PAAm) microparticles by on-chip photopolymerization from a W/O emulsion

Aqueous phase
100 g L^{-1} AAm
10 g L^{-1} BIS
5 g L^{-1} LAP*

Oil phase
2% (w/w) NH$_4$-salt of Krytox FSH (Dupont) in HFE 7500 (3M)

Total flow rate 1,200 μL h^{-1}

5 min. residence time
10 mW cm^{-2}

Squeezed droplets → Pancake microgels

*Lithium phenyl-2,4,6-trimethylbenzoylphosphinate

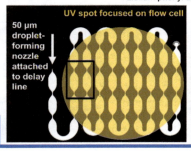

50 μm droplet-forming nozzle attached to delay line

Densely packed emulsion flowing through microchannel

Optimizing photoinitiator retention in aqueous droplets during UV-exposure

When moving from a continuous aqueous, reactive phase, as it would be used for forming bulk hydrogels, to a discontinuous one for forming hydrogel particles, the experimenter needs to keep in mind that the surfactant shell surrounding the droplets is a dynamic environment instead of a permanent barrier. Thus, a judicious microfluidics user should pay attention to the **retention of photoinitiators in the aqueous phase**. As a photoinitiator has been dissolved inside the aqueous pre-polymer solution, one should think that it remains in there throughout the UV-initiated droplet solidification.

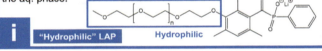

When exposed to UV light, **photoinitiators are usually decomposed into smaller reactive, e.g., more hydrophobic, building blocks**. Thus, it is crucial to optimize the surfactant concentration and to consider photoinitiators whose building blocks likewise show a sufficient solubility in the aq. phase.

Microfluidic experiment: Photopolymerization of microfluidic emulsions (Part 3)

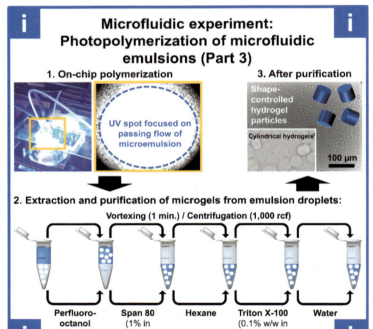

The protocol for forming microgels by on-chip UV polymerization introduced in the info boxes above can be easily modified toward individual needs of microfluidics users with respect to materials [39], polymerization triggers, as well as particle properties (in addition to geometry and size). For instance, instead of solidifying the microfluidic emulsion droplets by free radical polymerization using APS or LAP, polymerization or gelation can be based on UV-induced thiol–ene chemistry, and rather "boring" starting materials like AAm can be replaced by thermo-responsive ones (e.g., N-isopropylacrylamide, NIPAAm) that can serve as building blocks of switchable microbeads, capsules, valves, and vents. Furthermore, other than polymerizable monomers, a different class of microgel building blocks are prepolymerized macromolecular precursors that are gelled inside microfluidic droplets by polymer-chain interlinkage, which can again be achieved photochemically (e.g., by photodimerization of double-bond containing moieties such as maleimide or dimethylmaleimide side groups on the polymers) [40] or thermally (e.g., by coupling of two types of polymers, one carrying amine groups and the other carrying N-hydroxysuccinimide-ester groups).

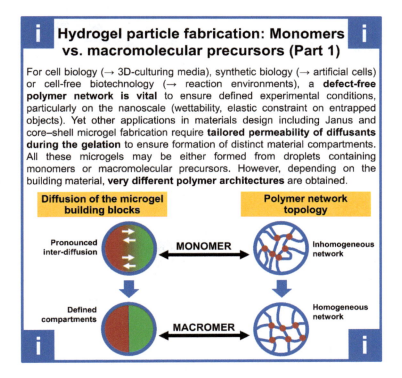

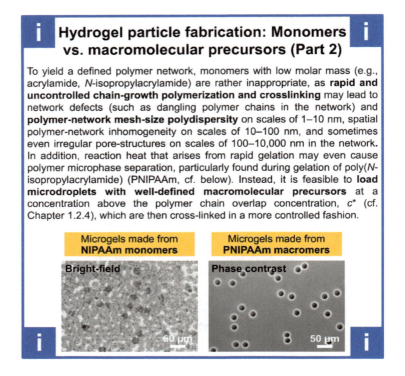

3.4.3 Beyond simple emulsion droplets: Double and higher-order emulsions

The latter section has shown us how easy to produce yet versatile microfluidic emulsion droplets are. So far, though, we have limited our discussion to the simple case of single-emulsion droplets, with a particular focus on the W/O variant. In the next step, we add a level of complexity and regard double-emulsion or even higher-order emulsion droplets. Double-emulsion droplets consist of a core–shell structure of an inner droplet nested inside an outer droplet that itself is dispersed in a continuous medium; with this morphology, they inherently resemble a microcapsule, which foreshadows their key area of utility. Higher-order emulsion droplets such as triple- or quadruple-emulsion droplets exhibit core–multi-shell structures and are therefore suitable for even more sophisticated microcapsule templating. Producing these droplets is actually straightforward. To make double emulsions, we simply need to add a second channel junction to perform a second emulsification step after or concurrent to a first. In glass-capillary microfluidics, this can either be achieved by two sequential capillary orifices in a co-flow device, or by two orifices frontally facing each other in a flow-focusing geometry, as shown in Figure 2.6 of Section 2.2.3. In PDMS-based stamped-cell microfluidics, this is achieved by two sequential channel cross-junctions operated such to have two dripping instabilities in either junction, as shown in Figure 2.7 of Section 2.2.3, or by having a co-flow situation in the first and a dripping situation in the second junction, as shown in Figure 2.9 of Section 2.2.4. In all these variants, tuning the flow-rate ratios serves to tune the overall droplet size and the shell thickness. Especially the variant of one-step double-emulsification shown in Figure 2.9 of Section 2.2.4 can serve to make extra thin-shelled double emulsions. In addition, the fluid flow rates also determine whether just one or a well-defined integer number of multiple inner droplets are encapsulated into each outer droplet.

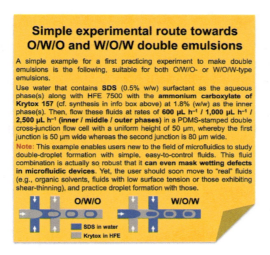

Simple experimental route towards O/W/O and W/O/W double emulsions

A simple example for a first practicing experiment to make double emulsions is the following, suitable for both O/W/O- or W/O/W-type emulsions.

Use water that contains **SDS** (0.5% w/w) surfactant as the aqueous phase(s) along with HFE 7500 with the **ammonium carboxylate of Krytox 157** (cf. synthesis in info box above) at 1.8% (w/w) as the inner phase(s). Then, flow these fluids at rates of **600 μL h⁻¹ / 1,000 μL h⁻¹ / 2,500 μL h⁻¹ (inner / middle / outer phases)** in a PDMS-stamped double cross-junction flow cell with a uniform height of 50 μm, whereby the first junction is 50 μm wide whereas the second junction is 80 μm wide.

Note: This example enables users new to the field of microfluidics to study double-droplet formation with simple, easy-to-control fluids. This fluid combination is actually so robust that it **can even mask wetting defects in microfluidic devices**. Yet, the user should soon move to "real" fluids (e.g., organic solvents, fluids with low surface tension or those exhibiting shear-thinning), and practice droplet formation with those.

3.4.4 Beyond simple aqueous droplets: Complex fluids in SF microfluidics

In Chapter 1, we have discussed the behavior of Newtonian versus non-Newtonian liquids – for instance, in view of the viscosity of such liquids under shear stress – and the consequences for their handling in microfluidic channels (e.g., for droplet formation). After establishing the general experimental details on emulsion formation from simple aqueous Newtonian solutions in Section 3.4.2 and 3.4.3, we now want to extend our discussion on droplet formation toward more complex liquids, either being "complex" due to their non-Newtonian flow behavior or because they are made of a delicate mixture of differently behaving compounds, or even because they contain particulate content. In the versatile realm of microfluidics-based biological applications, complex fluids typically concern reaction mixtures for PCR, cellular extracts like cell lysates, protein or enzyme solutions, or even living-cell suspensions in a serum fluid. In the realm of materials science, complex fluids typically concern non-Newtonian and highly viscous polymer solutions or colloidal suspensions, whose flow characteristics have been detailed earlier in Sections 1.2.4 and 1.2.5. With such complex fluids, controlled droplet breakup in a well-defined and practically realizable dripping regime is tough to achieve. To overcome this limitation, we need tools to enforce the droplet breakup. One tool to achieve that is to use one-step emulsification, as shown in Figure 2.9 of Section 2.2.4. In this approach, a hard-to-emulsify fluid is first surrounded by an easy-to-emulsify chaperone fluid in a first channel junction, whereupon these two fluids flow in a laminar parallel fashion alongside each other. In a second channel junction, droplet breakup of this co-flow, which has the easy-to-emulsify fluid at its sides, is performed by flow-focusing with an external fluid, thereby enforcing the inner hard-to-emulsify fluid to break up as well. Another approach to enforce droplet breakup is to use geometric constrictions in the microchannel that locally cause extra strong shear forces or to even employ auxiliaries such as external mechanical energy, for example, sound waves that hit the flowing fluid stream and induce its breakup.

When we want to process particulate content in an emulsion-forming microfluidic experiment, we need to distinguish between dilute particle suspensions, where the passage of a particle through a microchannel is merely occasionally happening, or dense particle suspensions up to concentrations where these suspensions behave more like a colloidal jam. In this context, note that the word "particle" is actually a place holder for any kind of micrometer-sized entity, be it a microgel, a granule, or even a living cell. Such entities may be subject to encapsulation into emulsion droplets for further downstream treatment, for instance, for rapid (bio)analytical screening or for further particle templating to permanently encapsulate the first particulate content. During the droplet encapsulation, the number of encapsulated entities per droplet will be Poisson-distributed in the case of a dilute suspension entering the microfluidic channel junction where the droplet breakup and content encapsulation occurs. By contrast, if a dense-packed jam of the particulate entities enters the device, their encapsulation statistics can beat the Poisson distribution. This is especially achieved if a device is constructed such that the particles enter the microchannel cross-junction in which they are encapsulated into droplets in a dense single-file flow, because then, a controlled integer number of them will enter each droplet, simply determined by the particulate-jam flow rate and the droplet-fluid as well as the continuous-fluid flow rate. To achieve such a single-file flow, the microchannel diameter must be equally sized or even slightly smaller than the particles; as long as they are soft and deformable, such as cells and microgels, this will enforce the particles to flow through these too-small channels in a well-defined single-file pattern, along with slight squeezing of the particles. Of course, the microchannel should not be made too small to avoid overly high back-pressure in this situation. An exemplary application of such highly defined particle encapsulation is the formation of polymer microgel scaffolds for controlling the emulsion shape toward nonequilibrium structures, as shown in Figure 3.42 [41].

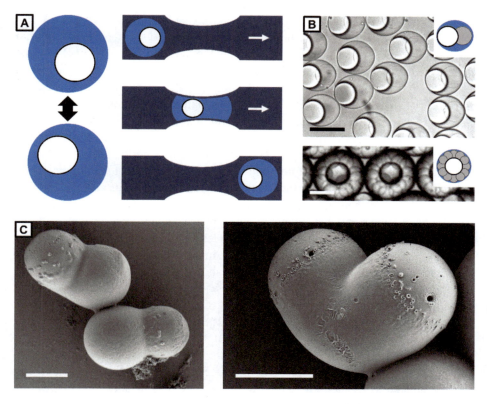

Figure 3.42: Particle scaffolding for controlling the shape of microfluidic emulsion droplets to be nonspherical. (A) Schematic of a double-emulsion droplet with random inner drop position (left) passing through a microchannel constriction (right). (B) Nonspherical (top) and spherical (bottom) double emulsion with controlled morphology by microgel particle scaffolding. The scale bars are 50 μm. (C) SEM images of nonspherical, anisotropic microgels with two (left) or three particle cores (right) templated by solidification of shape-arrested double-emulsion droplets like those shown in Panel A. Adapted from [41]; copyright 2011 Royal Society of Chemistry.

3.4.5 Formation of vesicles from double-emulsions templates

The prime area of application of core–shell double-emulsion droplets is templating microcapsules from them. The most simple approach to achieve that is to solidify the double-emulsion shell phase by polymerization or gelation, thereby solidly engulfing the inner droplet core. With that approach, however, the resulting capsule may actually turn out to be too strong for an application. This is because encapsulation usually targets at controlled delivery and release of the encapsulated ingredients, requiring the capsule shell to be switchable on demand such to allow the payload to be released from it. On top of that, it is a general desire to a microcapsule that it shall have a shell as thin as possible, such to have maximal inner core volume at a given overall core–shell size. Both these desires are fulfilled with **vesicles**, which are core–shell entities consisting of a just mono- or bimolecular layer of either amphiphiles or polymers in their shell. These entities may be templated from double-emulsion precursors if they are loaded with amphiphiles that will later form the vesicle shell layer.

Based on the theoretical background provided in Section 2.4.3, we can readily focus on the experimental aspects of vesicle formation from double-emulsion templates and outline a (PDMS-based) microfluidic experiment for vesicle formation from microdroplet-based templates. These templates need to consist of an aqueous core surrounded by a hydrophobic phase. Water-in-oil-in-water double emulsions directly reflect this exact architecture. Consequently, we require a microfluidic device consisting of two flow-focusing junctions, forming water-in-oil droplets at the first junction, which are then surrounded by an aqueous outer phase at the second junction (for generating a suitable microchannel surface wettability, cf. Sections 3.4.1 and 3.5; also note in this context: as most amphiphiles require organic solvents for dissolution, spatial control over the surface wettability should be based on a protective glass-like sol–gel coating). A suitable device is shown in Figure 3.43.

This device is fed with an aqueous phase at the first and third inflow port and the amphiphile dissolved in a mixture of at least two organic solvents at the middle inflow port. Why are the vesicle building blocks not simply dissolved in *one* organic solvent, but instead, in a mixture of *two* solvents? This is for two reasons. First, the organic solvent phase should ideally balance the density mismatch between the double-emulsion droplets and the surrounding aqueous phase after their collection (and before emulsion-to-vesicle transition). This allows the organic solvents to be evaporated off by diffusion through the surrounding aqueous phase (in which, for this purpose, the organic solvents must have a finite small solubility) at a steady and defined rate. A single organic solvent cannot fulfill these requirements. Second, use of an organic solvent mixture will cause change of the shell-phase solvent composition during the evaporation, as this occurs with different extent and rate for the different mixed organic solvents, thereby changing the polymer solubility state in the shell phase and entailing controlled polymer assembly to form a vesicle membrane. Commonly, mixtures of chloroform and hexane (premixed at a ratio of 80: 20 or 70: 30 v/v) or toluene and chloroform (premixed at a ratio of 65: 35 v/v) do a good job. Exemplarily, in the case of poly(styrene)-*b*-poly(ethylene glycol) (PS-*b*-PEG), microfluidics users should choose the former solvent mixture.[34]

Next, we need to identify a set of inner and continuous phases. As introduced in Section 2.4.3, the composition needs to account for the desired content of the later vesicle and prevent osmosis-driven shrinkage of the emulsion droplets during

---

[34] The actual amphiphile concentration needs to be based on the molar mass of the amphiphile and its interfacial occupancy. For further reading, se R. C. Hayward et al., *Langmuir* **2006**, *22*, 4457. **DOI: 10.1021/la060094b**.

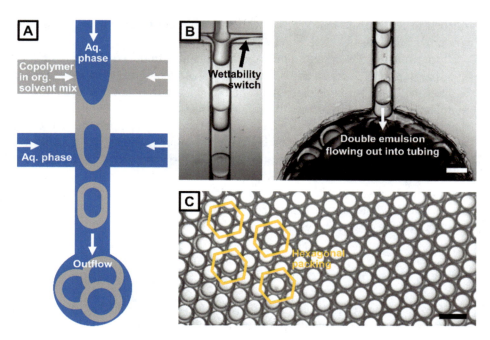

Figure 3.43: Microfluidic fabrication of copolymer-loaded double-emulsion droplets serving as templates for directed formation of giant polymersomes. (A) Schematic of a microflow cell with an injection pattern suited for forming copolymer-stabilized water-in-organic-solvents-in-water (W/O/W) double emulsions. The copolymer (here, PS-*block*-PEG) dissolved in the middle organic-solvent phase serves both as a surfactant stabilizing the double-emulsion droplets against coalescence and as a building block for the to-be-formed vesicles. (B) Bright-field microscopy images of double emulsions formed at a microchannel cross-junction and then flowing through the outflow channel of the microfluidic device. The scale bars are 100 μm. (C) Bright-field microscope image of collected double-emulsion droplets. Hexagonal packing (highlighted) indicates the perfect droplet-size monodispersity. The scale bar denotes 200 μm.

W/O/W emulsion-to-vesicle transition. For example, both the inner and continuous phase can be made of aqueous solutions of glucose (5 mM to 50 mM). Using a microfluidic device as sketched and shown in Figure 3.43A and B, typical flow rates are 200 μL h^{-1}, 400 μL h^{-1}, and 800 μL h^{-1} for the inner, middle, and continuous phase, respectively. As the amphiphile-stabilized double emulsions are formed (Figure 3.43C), the organic solvent phase will soon start to reduce its volume, enabled by partial miscibility of one or more of the organic solvents in the surrounding continuous aqueous phase. In this step, it is vital to maintain a constant evaporation rate. A too sudden evaporation may lead to rupture of the double-emulsion shell, whereas too slow evaporation will preserve the emulsion state and hamper a successful transition into vesicles, unnecessarily slowing down experimental progress. On this account, microscopy chambers consisting of an O-ring that is attached to a glass slide and sealed with a transparent tape[35] on the upper side have proven to be useful collection vessels.[36]

If we use a mixture of chloroform and hexane, as introduced above, the higher vapor pressure and better solubility in water causes chloroform to evaporate faster than hexane from the emulsion droplet shell. Thus, the emulsion will become density-mismatched with the surrounding aqueous media floating to the top of the collection chamber, while the inner phase volume, which determines the size of the later polymersomes, nearly remains constant. Eventually, chloroform and hexane are sufficiently evaporated such that the (PS-PEG-based) amphiphile monolayers at the inner–middle and the middle–outer interfaces assemble into a membrane, yielding the desired vesicles.[37] Note: depending on the amount of residual hexane inside the vesicle membrane, experimenters may simply skim off the vesicles near the surface of the collection vessel.

3.5 Trouble shooting in microfluidics

With the preceding insights into the various levels of microfluidics in practice, from design over fabrication to application, the successful realization of continuous flow or emulsion-droplet formation seems to be a rather simple task, at least if one is not "all thumbs." Yet, however, everyone who has had a glimpse into the scientific every-day business, irrespective of the specific field, may know one fundamental experience: more than 90% is failure. Yet, this is exactly how we proceed in science. Quoting Samuel Beckett: "*Ever tried. Ever failed. No matter. Try again. Fail again. Fail better.*" To make it easier for a microfluidics user to "fail better", the following final chapter of this textbook shares some of the authors' experience on common sources of error and how to circumvent or fix them.

---

[35] These tapes often have two holes that can be utilized to feed the outflow of the microflow cell in one hole, while excess air from the microscopy chambers is pushed out through the other hole.

[36] Eppendorf tubes that are commonly used as collection vials in microfluidic experiments should be avoided due the use of organic solvents. Moreover, a multi-layered assembly of emulsion droplets in these vials may lead to destabilization depending on the amphiphilic properties of the vesicle building blocks.

[37] While vesicles made by directed self-assembly based on emulsion templates are usually perceived to be unilamellar (as there are only two single amphiphile layers involved), it should be noted that it is rather challenging to identify the exact amphiphile concentration at which no excess molecules will be trapped between the as-formed vesicle bilayer. In fact, the amphiphile concentration can be utilized as a parameter to tailor the vesicle membrane thickness as received by the double-emulsion templates and to deliberately incorporate a defined amount of amphiphiles within the vesicle bilayer.

Let us start with the design of stamped microfluidic devices[38] based on CAD as the very first step in device fabrication based on combined PL and SL, as discussed previously in Section 3.1.2. Microflow cell design by CAD-based drawing tools such as Autodesk AutoCAD, SketchUp, or FreeCAD enables users to develop microchannel networks with complex architecture and function. In this context, users should not be scared by the vast design capabilities and options of CAD programs, as these have become easily accessible also for non-computer affine users, thanks to a wide range of (free-of-charge) training courses, simple written handbooks (e.g., "AutoCAD for Dummies" by B. Fane, ISBN: 978-1-119-25579-6), and online tutorials.

After successfully managing the CAD-based design of a microchannel layout, a crucial step remains: the transfer of the CAD drawing into the actual photomask that can be used for flow cell master fabrication by photolithography (Section 3.1.4). For that, a dxf- or dwg-file, which are common file types in CAD, needs to be converted into a photomask printer-compatible format. For users new to the field of microfluidics, it is generally suggested to utilize the service of commercial mask printing companies for file translation. Commercial vendors for photomask printing include JD Photo Data (UK), Photomask Portal (US), Fineline Imaging (US), or BlackHole Lab (FRA). These companies also usually offer a conversion service and subsequently send out a PDF check plot. With these PDF proofs, the to-be-printed photomasks can be screened for mask defects to assure correctness of the file conversion and printing process. On top of that, we recommend each photomask obtained from a commercial printer to be carefully checked with regards to surface roughness (Figure 3.44A) and microchannel resolution (Figure 3.44B). As mask printing services usually do not know about the purpose of the as-received digital file (showing a microchannel network, for instance), they will usually not point out any odd-looking flow cell features. Yet, simple light microscopy may reveal later that the resolution of the printing process may not match the desired minimal feature size as designed in CAD. Thus, structures that are not connected in the original CAD file may accidently become connected on the photomask or appear rough.

Commonly, commercial vendors charge customers for file conversion, and as microflow cell development is a continuous process toward optimal designs and experimental conditions, respectively, such costs, which can be tens of euros per order, can add up quickly. Therefore, it is worth considering conversion of CAD files oneself, as detailed in the info box on converting dfx files into Gerber files. When printing photomasks on a CAD printer on their own, microfluidics users may experience another common photomask error, which arises from the translation of the CAD file (usually *.dxf or *.dwg) into a printer-compatible file type, usually the Gerber-file type (cf. info box on converting dfx files into Gerber files). In the conversion process, the

---

[38] For trouble shooting on the manufacturing and operation of glass-capillary-based microfluidic devices, please continue at the end of this chapter.

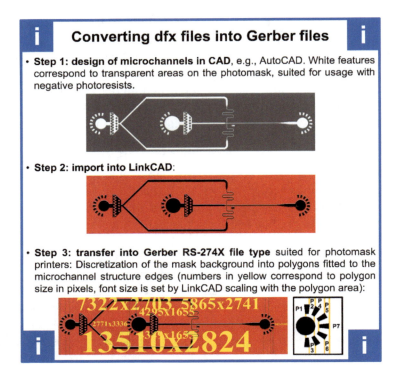

corners of any polygon in the CAD file will serve as a starting point for discretizing the mask into many subsections. These so-called polygons can be accidently disrupted by features of the microflow cell design, which manifests itself in the form of parallel lines added to the flow cell design, as exemplarily shown for diamond-type structures surrounding the round microchannel pad in Figure 3.44C.

With an optimized photomask at hand, the next step in the microfluidic device making procedure is its use for master device fabrication via photolithography. In this step, there is one key mistake to be avoided in the photomask handling. Photomasks are only printed on one side, whereas the other side of the foil is blank and thus flat. It is this flat side that needs to face the resin surface underneath when transferring the photomask channel structure into the photoresist. If the photomask is accidentally placed upside down, this will cause a gap between the photomask and the resin, in which we will lose sharp contrast during the UV illumination, thereby reducing the resolution of the channel structure on the photoresist, and as such, also in its PDMS replicas later. But how to identify the correct side of a photomask? This can be done by making a scratch with a razor blade in an area that is not crucial for later microfluidic device operation. The patterned side of the photomask will be harmed by the scratch in a sense that its black coating gets partially destroyed.

Another frequent error during the process of photolithography is delamination of the photoresist during the soft bake step. If that happens, we recommend to

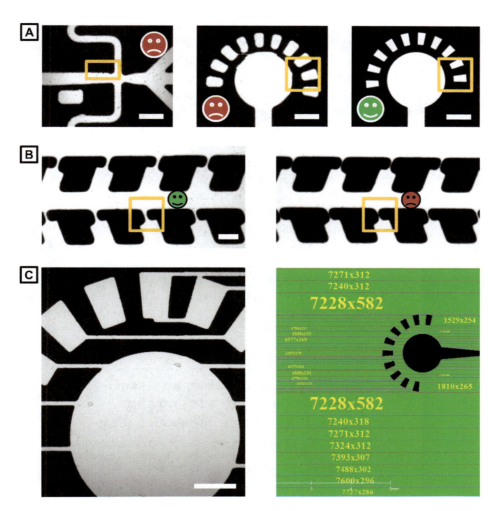

Figure 3.44: Printing defects in photomasks designated for microflow cell fabrication by photolithography employing negative photoresists, for example, SU-8 (MicroChem). Areas of interest are highlighted in Panel A and B. (A) Photomask with fringy edges (left, middle) compared to a photomask with desired sharp edges. The scale bar is 100 μm in the left and middle image and 500 μm in the right image. (B) Photomask with expected separate structures (left), and undesired jointed structures (right). The scale bar denotes 50 μm. (C) Photomask with additional horizontal lines, sometimes unwantedly appearing in the conversion process from a CAD drawing to a photomask printer-compatible file. The scale bar is 500 μm.

check if there is contamination of the fluids used for the cleaning of the carrier wafer before the resin spin coating. If this is not the cause for the error, then we recommend to use wafers with a higher surface roughness or to precoat the wafer with an adhesion promoter like TI Prime (Microchemicals GmbH), which optimizes the surface chemistry and contact angle for improved wetting of the photoresist.

The microchannel edge sharpness may not only suffer from incorrectly (i.e., upside-down) positioned photomasks during the photolithography, as discussed above, but also because of overexposure or underexposure, as shown in Figure 3.45. Furthermore, in the following hard-bake step, reduced device resolution can result from resin decomposition and cracks due to over-backing.

After successful master device fabrication, the most common source for microfluidic device rejects is insufficient bonding of the PDMS replica to the glass substrate, leading to leakage during microfluidic experiments, even at low fluid pressure. Plasma bonding of PDMS and glass seems to be a simple process, but it actually depends on many parameters. The info boxes in the following provide experimental details that can help to prevent common errors leading to insufficient bonding.

In simple cases of droplet-based microfluidics, it is sufficient to treat the channel-wall surface with a suitable silane reagent to render it hydrophilic or hydrophobic. The latter may also be achieved by "intended misuse" of the commercial car

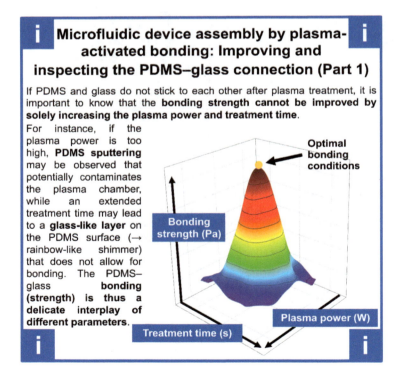

i Microfluidic device assembly by plasma-activated bonding: Improving and inspecting the PDMS–glass connection (Part 1) i

If PDMS and glass do not stick to each other after plasma treatment, it is important to know that the **bonding strength cannot be improved by solely increasing the plasma power and treatment time**. For instance, if the plasma power is too high, **PDMS sputtering** may be observed that potentially contaminates the plasma chamber, while an extended treatment time may lead to a **glass-like layer** on the PDMS surface (→ rainbow-like shimmer) that does not allow for bonding. The PDMS–glass bonding (strength) is thus a delicate interplay of different parameters.

Microfluidic device assembly by plasma-activated bonding: Improving and inspecting the PDMS–glass connection (Part 2)

Beyond the effect of plasma power and treatment time (Part 1), other parameters that need to be considered when optimizing the PDMS–glass bonding include **surface (e.g., finger prints)** and **chamber (e.g., dust) contamination**, **gas composition (e.g., humidity**, which even may depend on the season of the year when using "normal" air instead of bottled gas with tailored composition), **post-plasma treatment delay** (leading to a decrease in surface activation), and **choice of gas** (e.g., N_2, O_2, air). With this large parameter space in mind, it is crucial to identify simple methods that allow for efficiently monitoring the PDMS–glass bonding:

Contact angle
Put a **water droplet onto the freshly activated surface** (PDMS / glass). The **contact angle should be ~ 20°**, which corresponds to a **bonding strength of 0.2 to 0.3 MPa**.

Peel-off behavior
Destructive, but informative: try to remove the PDMS replica from its substrate. The **chip should be ripped apart** with PDMS residues remaining on the glass surface indicating optimal bonding.

Microfluidic device assembly by plasma-activated bonding: Improving and inspecting the PDMS–glass connection (Part 3)

To inspect the quality of bonding between glass and PDMS, two more characterization methods may be considered:

Pressure resistance
Connect the assembled flow cell to a syringe pump and flush the microfluidic device at high pressure, e.g., with **flow rates 5 to 10 times higher than in the desired experiment**. No leakage should be observed.

Spreading of bonding
When gently putting a PDMS replica onto a glass slide right after activation, the **spreading front** (contact line between air–PDMS–glass and PDMS–glass) should proceed quickly and in all areas of the flow cell **without (!) additional mechanical (manual) compression**. Just the force exerted by the weight of the PDMS replica onto the glass substrate should be sufficient. To follow the spreading front, note that bonded areas appear darker than un-bonded areas.

NOTE: While additional **manual squeezing of a PDMS replica and a glass substrate** does not change the surface chemistry (relying on the conversion of Si-OH groups into Si-O-Si connections), it may still help to drive out air entrapped between PDMS and glass.

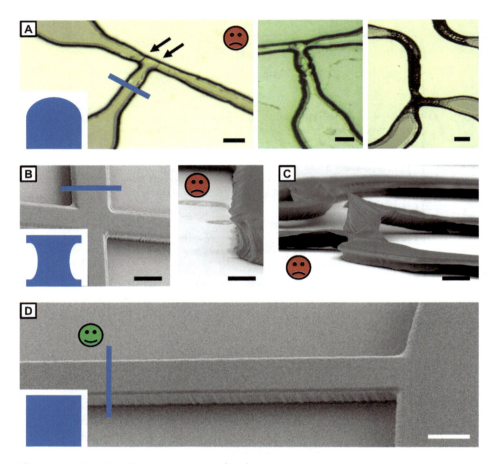

Figure 3.45: Scanning-electron microscopy (SEM) characterization of common defects in microchannel master devices made from a SU-8 resin by photolithography. (A) Microchannels with round edges on their top due to accidental scattering of UV light between the photomask and the resin layer during UV-light exposure as well as due to overly extensive washing of the master device in the developer bath. Additional cracks in the final microchannel structure (highlighted by arrows) indicate overly extensive hard baking in the final fabrication step. The scale bars for all panels denote 50 µm. (B) Microchannels exhibiting walls that are curved inwards due to insufficient UV-light exposure of the SU-8 resin, followed by excessive removal of the unpolymerized resin. The scale bar is 50 µm and 10 µm for the zoom-in picture on the right. (C) Similar to (B), insufficient UV-light exposure of the SU-8 layer leads to incomplete polymerization of the microchannels at the bottom, and thus delamination during the hard-baking step. The scale bar denotes 50 µm. (D) Microchannel structure without defects for comparison. The wavy profile of the microchannel walls arises from the limited resolution of the lithography mask (cf. Figure 3.2). The dark layer at the bottom of the microchannels is an optical effect due to the limited depth of field of secondary electrons in the SEM characterization. The scale bar is 50 µm.

windshield care product Aquapel (Pittsburgh Glass Works LLC). Hydrophilicity, in turn, can also be generated by quick plasma treatment. Whereas this is all simple, we encounter a more complex scenario if we intend to use our device with either volatile organic solvents such as hexane, toluene, or chloroform, because these all swell PDMS and thereby cause channel-diameter constriction up to complete clogging, and/or if we intend to make double emulsions and therefore require spatially patterned channel-wall wettability. For all these purposes, we must apply a sol–gel coating to our channels. This treatment is prone to cause several sorts of trouble in microfluidic experiments.

The most obvious error that we may cause by sol–gel coating our channel walls is to **clog** them if the coating is too thick. This is particularly dangerous in regions of the microchannel network where we have small passages for the fluids, for example, in the regions between the diamond-shaped pillars of dust filter units right behind the fluid inlets, as seen in Figure 3.46A. In the contrary case, a too-thin sol–gel coating will come along with insufficient function of the coating, that is, insufficient channel stabilization against swelling and insufficient ability for spatially resolved channel-wall wettability treatment, as seen in Figure 3.46B. From the authors' experience, in general, the device-production step of applying a sol–gel coating causes about 50% of the channels on a chip to be discarded later, as the sol–gel layer will turn out to be not good enough on them. The other 50% may be subject to either direct use or further treatment, for example, to spatially resolved wettability treatment.

The most common source of error in the step of channel-surface grafting to achieve wettability treatment is **insufficient polymerization**, as it typically occurs if the monomer or the (photo)initiator is too old and therefore have lost reactivity

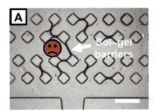

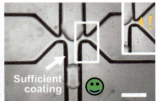

Figure 3.46: Wrong and right application of sol–gel coatings on microchannel walls. (A) Bright-field microscopy image of a filter unit with diamond-like posts at the beginning of a microchannel. The sol–gel coating recipe utilized in this example has led to a thick layer on the microchannel surface that partially blocks the filter posts. (B) Microchannel cross-junction with a sol–gel coating for the production of water-in-organic-solvent droplets. An insufficient coating thickness, as shown in the left image of Panel B, may keep the microchannels open enough for operation, but on the other hand, the surface wettability will then not be distinct enough to enable W/O droplet formation. In contrast, a thick sol–gel coating, as shown in the right image of Panel B, will entail correct and sufficient wettability for W/O droplet formation, but in this case, small features within the flow cell may become critically narrow, which may then promote clogging either by debris, dust, or by the sol–gel itself during the coating step, especially if we consider the inherent experimental variation from device to device. The scale bars for (A) and (B) both denote 100 μm.

over time. Users may get a notion of that if the typical polymer-coating structure on the channel surface, as seen in the respective info boxes in Section 3.4.1, does not appear. On the contrary, **overly strong polymerization** will cause this polymer layer to penetrate into the PDMS material base, leading to swelling and wiggling of it, which then causes delamination of the sol–gel coating from the channel walls. This is seen in Figure 3.47. If a very-fresh and reactive monomer is used, it may also happen that it already polymerizes in its supply syringe by thermal auto-initiation.

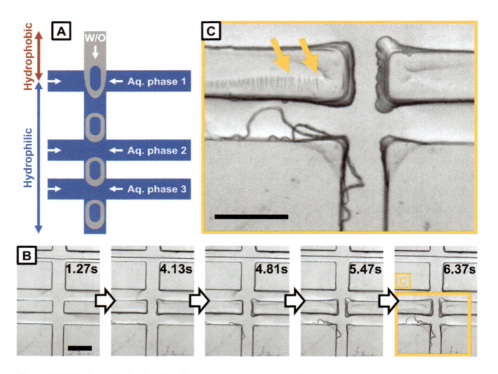

Figure 3.47: Sol–gel delamination from a microchannel surface during graft polymerization. (A) Sketch of a to-be-coated microfluidic device for forming W/O/W double emulsions. In this device, three sets of inflow channels (aq. phase 1–3) allow for further downstream addition of compounds to subject the double-emulsion droplets to further treatment, for example, additional surfactants to further stabilize them or polymerization initiators to solidify them. (B) Bright-field microscopy image sequence of the grafting process using the example of grafting-to polymerization of poly(acrylic acid) (PAA) spatially localized by flow confinement to the lower part of the microfluidic device. The liquid–liquid interface of the blocker and the reactive phase are visible in the uppermost cross-junction. (C) Enlarged section of the microfluidic device section from (B). Arrows indicate the penetration of PAA pass the functional sol–gel coating into the PDMS matrix, leading to swelling, which then causes the sol–gel coating to delaminate from the microchannel surface as fiber-like debris. The scale bar for all panels in (B) and (C) denotes 200 µm.

The wettability-treatment step has a similar failure-to-success ratio as the preceding step of sol–gel coating, such that in the end, about a quarter of the initially made channels work well. This may seem a rather high loss, which it indeed is, but at least the good quarter of all channels can be used multiple times if they are flushed and rinsed after each experiment. The only great challenge lies in parallelization of many devices, as it is needed in numbering-up of microfluidics for potential industrial use, because then, a much higher success rate must be achieved. Based on the above practical experience, the authors see sol–gel coating and channel surface grafting for wettability treatment to be improper for such numbering up, but still perfectly practical on common laboratory scales. In a view of numbering up, it appears more suitable to use channels with multiple heights, in which the channel-wall wettability does not play that great role as it does in uniform-height channels. To further enhance their performance, such multiple-height devices can be made globally solvent resistant by vapor-deposition of a protective layer of **parylene**.

Once we have a good and reliably operable microfluidic device at hand, we can start to use it. During that, another number of potential errors can happen. Rather often, microfluidic experiments are performed with molecular solutions that behave as Newtonian liquids. These are easy to process, and flow cells can be operated in a continuous fashion over many hours with them. However, as soon as we move away from these simple fluids toward complex solutions with high viscosity or insoluble content that is transported inside these fluids, experimenters face a variety of challenges. The non-Newtonian nature of many polymer and colloidal solutions and suspensions is inherent to them, and users will have to figure out for each specific case at hand in what range of flow conditions a desired flow pattern (such as controlled dripping in SF microfluidics) can be achieved. It may be helpful to also apply auxiliaries such as enforced droplet breakup, either by a chaperone fluid in a one-step process of double-emulsification, as discussed above in Section 3.4.4, or by additional means such as sound waves that destabilize a to-be-broken fluid jet. A particularly challenging case is if our fluids aren't solutions but suspension of above-colloidal-scale entities such as above-micrometer scale microgels or living cells. If these objects are rather stiff and cannot be easily flushed through a microchannel, clogging may be a key challenge.

If microchannel clogging persistently occurs, we first need to identify the reason for it. This can be experiment-related or device fabrication-related. Reasons for the latter case, and ways around them, have been discussed just before in the preceding text section. In the former case, clogging is caused by precipitation, agglomeration, and microchannel-wall adhesion. For example, oversaturated solutions of salts or polymers can quickly evaporate inside the microchannels due to their large surface-to-volume ratio and then precipitate. On top of that, when we use suspensions of above-colloidal vesicles, micelles, tubes, wires, or cells, incompatible size, high aspect ratio, or limited elasticity may hinder their free passage through the microchannel network. In that case, the microfluidic device design must account for the physicochemical and mechanical properties of the fluids and particulate contents to be used and tailor the channels such to allow for a free passage of these objects through them in the first place.

While clogging and delamination is an issue of any microfluidic device operation, there are further distinct challenges associated with CF and SF microfluidics. In the former case, a prominent challenge is continuous operation of nucleation and self-assembly processes at liquid–liquid interfaces (cf. Section 3.3). Here, locally constant reaction conditions may lead to likewise local accumulation of material, as shown in Figure 3.48. A solution to this problem is to not use planar and rectangular cross-section microchannels, where the upper and lower microchannel wall is in permanent contact with the inner phase of a flow-focused fluid jet, but instead, to manufacture and use flow cells with nonplanar microchannels such as glass-microcapillaries that generally exhibit a coaxial flow profile and totally prevent contact of an inner jet to the channel walls, as it will be fully surrounded by the flow-focusing outer fluid.

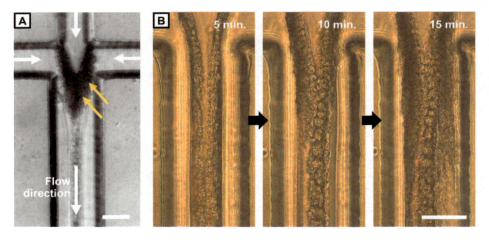

Figure 3.48: Flow obstruction and eventual device failure due to surface fouling of microchannels in the example of microfluidic polymer particle preparation by microprecipitation, as shown in Figure 3.29 already. (A) Bright-field microscopy image of a corresponding microfluidic co-flow experiment. A solution of polylactic acid (PLA, M_N = 5,000 g mol^{-1}) in acetone (1% w/w) is flow-focused into a thin jet by water to induce precipitation of polymer particles at the co-flowing liquid–liquid interfaces. As the solvent quality drops even at the microchannel cross-junction, PLA precipitates into polymer particles, which partially adhere to the microchannel walls and agglomerate there (highlighted by orange arrows), eventually filling out the whole cross-section of the outflow microchannel. (B) Image sequence of the microchannel surface fouling due to adherence of PLA to it. Initial agglomerate formation at the liquid–liquid interface quickly extends into the outflow channel and covers the whole lower and upper microchannel wall after 15 min. At this point, the flow profile of the flow-focused acetone jet is not centered anymore, but meanders through the distinct PLA agglomerates. The scale bars in (A) and (B) denotes 25 µm.

Related to the latter example of content-*ads*orption on the channel walls, another issue in microfluidic-device operation can be *ab*sorption of content inside the device material, that is, inside the PDMS elastomer network. A common example is "glowing microfluidic devices," which occurs when we use PDMS-based flow cells with fluids that contain fluorescent dyes. These dyes may penetrate into the PDMS elastomer or at least into its surface coating and cause it to fluoresce, as seen in Figure 3.49A. In general, microchannel surface fouling due to adhesion and ultimately penetration of small solutes into the flow cell is an issue; it is in fact one of the major restraints in commercialization of PDMS-based microfluidics. A partial solution to this problem is to apply a sol–gel coating to the channel walls to provide a diffusion barrier, or as a

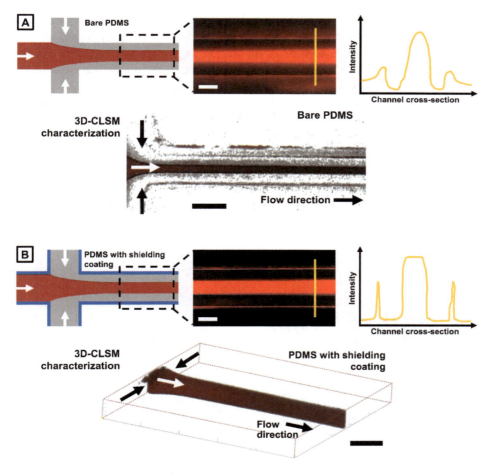

Figure 3.49: Diffusion of small-molecule solutes into the microchannel walls during CF microfluidic experiments in (A) a conventional, and (B) a glass-coated PDMS replica. In both examples, a solution of rhodamine is narrowed into a jet by plain water (left); corresponding line scans (right). The scale bar in both panels is 25 μm.

variant, PEG or BSA coatings to push back biofouling. Other than that, the only way to circumvent trouble with fouled channel surfaces is to apply quality control to the channels, that is, to periodically monitor their micro-optical appearance and operability in test experiments, and if permanent fouling and diminished device performance becomes evident, to discard and replace them by new ones.

One of the most severe negative effects of microchannel surface fouling is **alteration of the channel-wall wettability**. This will inherently happen if we use amphiphiles in a flowing fluid, as these easily attach themselves to the channel walls with their respective evenly philic side and have their oppositely philic side sticking out into the channel interior then, thereby exactly reversing the (local) channel-wall wettability.

In general, if the wall wettability is not well-defined, the performance of a narrow microchannel, in which an inner fluid stream can easily reach and touch the wall, can be drastically lowered. This is not so much an issue in glass-capillary microfluidic devices, whose channel walls are so far away from an inner-flowing phase that it can hardly have contact to them, but in PDMS-based stamped microflow cells, in which the channels are usually just as wide as the to-be-handled fluid streams or entities, this is a huge issue. Figure 3.50 shows how severe an **incorrect channel-wall wettability** can be in a SF experiment in a PDMS-based flow cell. In Panel A, we see a device with hydrophobically modified channel walls, correctly operating in a dripping mode and forming W/O droplets if it is fed with the to-be-inner water phase through its central inlet and the to-be-outer oil phase through its two side inlets. By contrast, if the channel-wall wettability is inverted, the same device and fluid-entering pattern does not operate in a defined dripping mode that forms W/O droplets, but instead, the designated inner phase will climb to the walls and wet them, whereas the designated outer phase wants to avoid wall contact and thus gets dispersed to O/W droplets carried by the water phase.

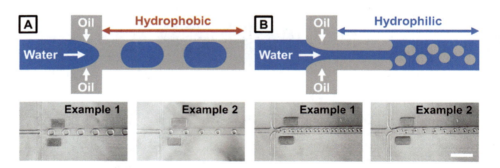

Figure 3.50: Water-in-oil emulsion formation in a planar flow-focusing junction with a hydrophobic outlet channel (left), in contrast to a hydrophilic outlet channel (right). For correct emulsion formation in the controlled dripping regime, the continuous (oil) phase needs to efficiently wet the microchannel walls, as it does in Panel A, whereas the opposite will happen if the channel-wall wettability is inverted, as seen in Panel B. The scale bar applies to all panels and denotes 250 µm.

A similar situation is encountered in co-flow experiments. Figure 3.51 shows how a to-be-centered fluid stream gets displaced toward wall contact if the wall wettability favors contact with it. A yet similar scenario can also be observed in double-emulsion formation, as seen in Figure 3.52. In this whole context, note that in general, simple fluid combinations (e.g., water and fluorinated oils) are more "forgiving" and may even mask microchannel surface defects. Thus, experimenters should test their devices not only with these easy candidates, but also with tougher fluids such as those needed in an actual application, for example, water–chloroform/toluene–water, which often serve in double-emulsion templated vesicle formation.

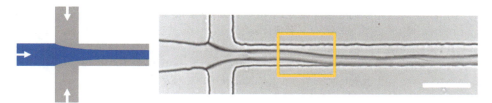

Figure 3.51: Defect wetting in microflow cells and its consequences for flow pattern formation in continuous-flow experiments. The micrograph shows an off-centered stream of a hydrophobic polymer solution in a microchannel with intermediate, ill-defined wettability. The scale bar for both panels denotes 50 μm.

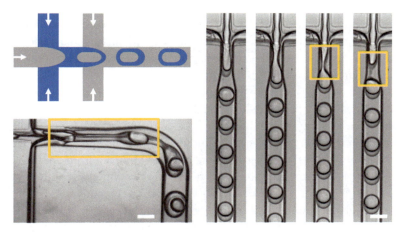

Figure 3.52: Defect wetting in microflow cells and its consequences for the flow pattern formation in segmented-flow experiments. The micrographs show the formation of W/O/W double emulsions (left) as well as a high-speed imaging sequence of the droplet break-off in the case of one-step W/O/W emulsion formation (right). Due to insufficient grafting of hydrophilic poly(acrylic acid) onto the walls of the flow cells' output channels, the droplet break-off is impaired and can only be enforced by additional features, such as a change of the microchannel geometry (highlighted), here in the form of a channel bend (left) or a channel-width constriction (right), in which high local shear forces act on the double emulsion's middle-phase flow. The scale bar is 100 μm in both panels.

Even without the specificity of wetting-defects and channel fouling, an inherent hurdle in any microfluidic experiment is its start-up. This is particularly puzzling in SF experiments. At first, we may have issues with undesired fluid backflow into regions of our flow cell where it is not supposed to be, for example, entrance of a to-be-centered inner phase into the inlet regions of the outer phase or vice versa, where uncontrolled emulsion formation will occur. Once all this initially formed ill-defined emulsion is flushed out and each fluid is injected only in the way as it is supposed to be, we must bring our device into a controlled dripping regime if we didn't have a lucky shot of starting conditions in the first place.

For this purpose, we can tune the flow-rate ratio (FRR) and the total flow rate (TFR), and both these parameters affect the presence of controlled dripping vs. uncontrolled jetting. In general, it is desirable to work at high FRR and low TFR for dripping, as shown systematically in Figure 3.53. For each given case at hand, the experimenter may play with these parameters to see how the FFR must be tuned to come to a desired droplet size, and how far the TFR can be increased to operate at good overall productivity while still being in the dripping regime.

If we have successfully overcome all potential sources of error in the process of microfluidic device fabrication and operation, the last potential challenges may arise in the variant of droplet-based microfluidics by insufficient stability of the resulting emulsion droplets, and in potential difficulties in their further use. Let us focus on the first aspect first. A natural necessity for sufficient emulsion stabilization is to fully cover each droplet with surfactant. To swiftly screen the influence of the surfactant architecture (e.g., AB, ABA, ABC block sequence) and concentration to fulfil that necessity, and with that, to achieve emulsion stability, a sharp needle can be pushed onto an as-collected emulsion sample on a glass slide that is mounted to a light microscope, as shown in Figure 3.54.

While the actual force that is required for emulsion destabilization by manual needle indentation is usually not defined, Figure 3.54 still provides a visual impression of the acting forces. As the needle tip indents the emulsion sample, the flexible glass slide bows under the exerted pressure, as indicated by the droplets that move out of focus. Under these experimental conditions, sufficiently stabilized emulsion droplets will be strongly squeezed between the needle tip and the glass slide, but they will not break or coalesce, as seen again in Figure 3.55A. By contrast, insufficiently stabilized emulsion droplets will coalesce between the needle tip and the glass slide, as seen in Figure 3.55B. While this droplet coalescence mostly propagates through the sample in a continuous fashion (Figure 3.55C), shape relaxation of a locally merged droplet population may also trigger coalescence of droplets that are even far away from the actual merging event (Figure 3.55D) [42].

To close our discussion on microfluidic trouble shooting, we shed a view on one of the key applications of water-in-oil emulsification by microfluidics, which is the templating of microscopic hydrogels from the droplets. In this application, there's two common sources of error. First, use of reactive hydrogel precursors such as very

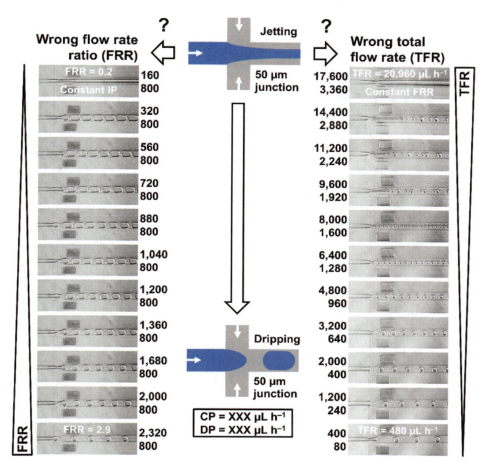

Figure 3.53: Search for good flow conditions in a flow-focusing water (blue) in oil (gray) (W/O) emulsion formation experiment by systematic variation of the flow rates of the dispersed phase (DP) and the continuous phase (CP), thereby moving the state of operation of the microfluidic device from a jetting to a desired dripping regime. The numbers next to each image denote the flow rates the outer fluid (upper number) and the inner fluid (lower number) in units of µL h^{-1}. At a high flow-rate ratio of the outer:inner flow (denoted FRR) and at a low total flow rate (denoted TFR), the microfluidic experiment transitions from an uncontrolled jetting state into a controlled dripping state. The scale bar is 200 µm.

reactive acrylic-type monomers (e.g., acrylamide) can lead to significant polymerization in the bulk of the monomer solution in its supply syringe already, thereby causing this fluid to increase in viscosity, so that at some point, controlled dripping can no longer be realized. Figure 3.56A shows an example of such a state. A way to prevent that is to cool the storage syringe by ice packs or by even operating the whole microfluidic experiment at low temperature (if possible even in a cold room), as shown

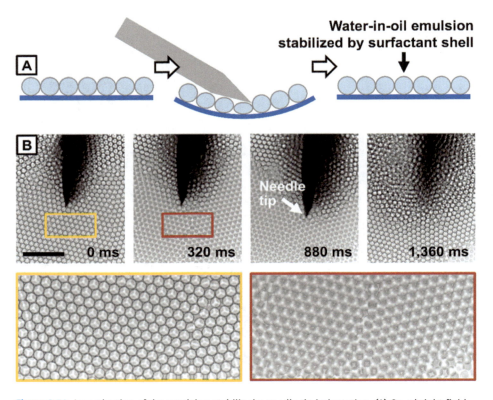

Figure 3.54: Investigation of the emulsion stability by needle tip indentation. (A) On a bright-field microscope, a commercial needle, 25G in size or similar, is brought into contact with a freshly prepared emulsion sample on a glass slide simply by hand or – if more precision is desired – utilizing a probe station with integrated micromanipulator. The emulsion stability is recorded at high speed (approx. 100 fps). Here, the sample is in focus during the initial stage of probing, and out of focus at a later stage, as the needle dents the glass slide.

in Figure 3.56B. Further help is provided if the monomer solution is not premixed with an initiator, but injected separate from it, such that contact of the monomer and initiator is delayed to happen on the device itself, right before droplet formation. For this purpose, we must use at least two channel cross-junctions, one for the initiator addition to the monomer solution (in a CF flow-focusing type) and one for the droplet-forming flow-focusing by addition of an immiscible outer phase. In contrast to this problem, other monomers may be too unreactive to form microgels with sound reproducibility, especially as the environmental conditions such as the lab temperature and the quality of the chemicals used may change from time to time or over time (for example, depending on the season). One example is the monomer N-isopropylacrylamide (that is famous for its thermo-responsive polymer in water), which is comparably unreactive, and on top of that, usually delivered in a stabilized from. In this case,

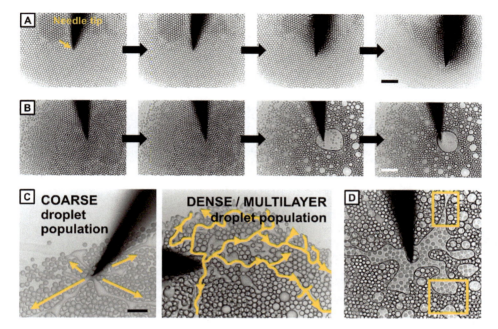

Figure 3.55: Typical emulsion behavior observed during needle indentation experiments. (A) W/O emulsion droplets stabilized by a surfactant shell. These droplets evade the needle tip or are squeezed between the needle and the glass slide. (B) W/O emulsion droplets stabilized by a surfactant with insufficient W/O interfacial coverage. As the needle indents into the emulsion sample, droplets spontaneously coalesce in an uncontrolled fashion, as indicated by increasing droplet-size polydispersity. The acquisition time for both image sequences in (A) and (B) is 2.5 ms. (C) Cascade-like coalescence among emulsion droplets originating from the needle tip-to-emulsion contact point. Depending on the droplet density in the emulsion sample, the resulting coalescence is spatially localized (left) or extends throughout the emulsion sample in a cascade-like fashion. (D) Shear forces induced by coalescing emulsion droplets that collapse and retract volume in one area of the emulsion sample may even induce emulsion breakage in another, spatially separated area. Scale bars are 250 µm in (A) and (B), and 100 µm in (C) and (D).

experimenters must spend more effort to assure reproducible experimental conditions along with taking means to assure sufficiently high reactivity, including adjustment of a reproducible high-enough temperature (yet not above or even too close to the LCST of the forming polymer, of course) and initiator content, use of re-crystallized monomer to remove stabilizers from it, and use of not too-old reagents in general.

A powerful way to circumvent the issues encountered in microgel formation by monomer polymerization, as just discussed above, is to use prepolymerized precursors that are not further polymerized but only crosslinked inside microfluidic pre-microgel droplets [40]. This reaction is often much more straightforward and may even lead to better defined network architectures than free-radical polymerization of monomers and crosslinkers, which often in fact entails quite marked network topological

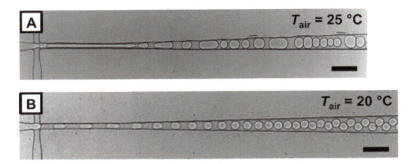

Figure 3.56: Influence of the laboratory air temperature on microfluidic experiments for forming polymer microparticles from premixed reactive monomer solutions. Exemplarily, poly(acrylamide) (PAAm) particles are microfluidically prepared from an aqueous solution of 60 g L^{-1} acrylamide (AAm), 1.8 g L^{-1} N,N'-methylenebisacrylamide (BIS), 8.8 mmol L^{-1} ammonium persulfate (APS), injected at 250 µL h^{-1}, and compartmentalized into droplets by a continuous phase composed of fluorinated oil (Fluorinert™ FC-40) with the ammonium salt of Krytox® 157 FSL (2% w/w) as a surfactant, injected at 800 µL h^{-1}. (A) At 25 °C, the premixed monomer solution starts to polymerize in the bulk of its reservoir in the supply syringe, thereby causing an increase in viscosity and formation of a liquid thread extending beyond the droplet-forming nozzle. (B) At 20 °C, by contrast, the premixed monomer solution does not yet markedly polymerize during the time of the microfluidic experiment, thereby allowing it to be compartmentalized into monodisperse droplets at the microfluidic cross-junction in a controlled fashion over the whole time of the experiment. The scale bars for both panels are 100 µm.

inhomogeneity. Yet, in the variant of polymer crosslinking instead of monomer polymerization, we inherently operate with higher viscous and potentially non-Newtonian fluids, which may be harder to emulsify than low-viscous and Newtonian monomer solutions. On top of that, the large size of the polymeric hydrogel building blocks and the high viscosity of their solutions markedly impairs their diffusive intermixing in microfluidic droplets, thereby prolonging the time to achieve droplet-interior homogenization. If the polymer crosslinking occurs before this homogenization is complete, then very inhomogeneous microgel particles will be obtained, as shown in Figure 3.57A. A way to minimize this undesired effect is to use polymer crosslinking reactions that are not overly fast and to enhance the droplet-content mixing by employing meandered and diameter-changing delay channels, as shown in Figure 3.57B.

To this point, our whole discussion of common sources of error and trouble shooting in microfluidics (specifically in its droplet-based variant) was focused on stamped flow cells in PDMS. We now spend another word or two on common trouble in glass microcapillary devices. Fortunately, there is much less to discuss for these devices, as they are generally much more robust and less prone to error than PDMS-based flow cells. This is for two reasons. First, glass is a hard, solid material tolerant to almost all fluids one could ever think about using in microfluidics, including volatile organic media and even somewhat aggressive acids, bases, or oxidants. All these cannot be used in PDMS as they would swell or even attack it, but glass is pretty

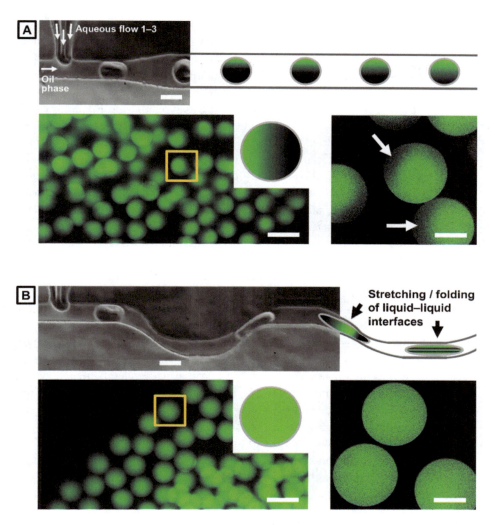

Figure 3.57: Preparation of polymer particles from a quickly reacting prepolymer solution and a crosslinker in microfluidic T-junction devices. The T-junction contains three inflow channels, the left and right one for the prepolymer solution and the crosslinker, respectively, separated by a central aqueous liquid stream acting as diffusion barrier for the prepolymers and the crosslinker. A fourth inflow channel coming from the left serves to compartmentalize the three-phase aqueous stream into monodisperse droplets. To visualize the interface of the nearly density-matched aqueous phases, differential interference contrast as well as fluorescence microscopy is employed. (A) Mixing of these fluids inside the droplets is incomplete, as indicated by the heterogeneous interior of the resulting microgels. (B) Mixing is improved by passing the emulsion through a periodically winding, meander-like outflow channel, whose diameter is slightly smaller than the droplet diameter. The resulting microgels exhibit a much more homogeneous material distribution than those from the experiment in Panel A.

resistant against them. Second, the perfect coaxial alignment of glass tubes in microcapillary devices along with the rather huge diameter of the outermost tube allows flow-focusing of fluids, either in a co-flow variant or in a droplet-forming variant, to be performed such that the inner fluid is fully surrounded by the outer, and that it has a large distance to the channel walls. These two circumstances make wettability adjustment much less necessary in glass-capillary devices than in PDMS flow cells. Based on all these features, there is actually just three limitations and potential sources or trouble in these devices. First, due to their manual assembly from a rather limited variety of building blocks (that is, commercially available square- and round-cross-section glass tubes), glass-capillary devices cannot be built with such great channel-structural variety as PDMS-based flow cells; they can also not be numbered up in a way as lithography allows for PDMS devices. Second, a usual source of error in the manufacture of glass-capillary devices is accidental clogging of the fluid inlets by epoxy glue that finds its way into the syringe-tip inflow ports too extensively and reaches the tiny capillary entrances underneath. A way to prevent that is to place the to-be-glued syringe tip right in the middle over the to-be-fed capillary entrance such that potentially entering fluid epoxy glue has to pass a long distance before reaching it, so there is a chance that it hardens before. On top of that, it is generally recommendable to glue these parts of the device not with freshly mixed epoxy glue, but with such that is some minutes old already and close to solidification. This epoxy mixture has pre-reacted already and thereby gained in viscosity, which decreases the risk of too-far flow underneath the syringe-tip inlets; also, such a mixture will harden soon, so even if it flows into clog-risky regions of the capillary arrangement, it may stop by solidification before actually causing clogging. As the third and last significant source of error, we name the risk of softening or even start of dissolution of the plastic parts of the fluid-inlet syringe tips in organic media that may be used in a microfluidic experiment. If this turns out to be an issue, then other sorts of fluid inlets should be used, for example, syringe tip-analogue pieces based on ceramics or metal.

References

[1] A. Waldbaur, H. Rapp, K. Länge, B. E. Rapp, "Let there be chip – towards rapid prototyping of microfluidic devices: one-step manufacturing processes", *Anal. Methods* 2011, *3*, 2681. **DOI: 10.1039/c1ay05253e.**

[2] I. Gibson, D. Rosen, B. Stucker, "Additive Manufacturing Technologies", Springer New York 2015. **DOI: 10.1007/978-1-4939-2113-3.**

[3] A. K. Au, W. Huynh, L. F. Horowitz, A. Folch, "3D-printed microfluidics", *Angew. Chem. Int. Ed.* 2016, *55*, 3862. **DOI: 10.1002/anie.201504382.**

[4] N. Bhattacharjee, A. Urrios, S. Kang, A. Folch, "The upcoming 3D-printing revolution in microfluidics", *Lab Chip* 2016, *16*, 1720. **DOI: 10.1039/c6lc00163g.**

[5] S. Waheed, J. M. Cabot, N. P. Macdonald, T. Lewis, R. M. Guijt, B. Paull, M. C. Breadmore, "3D printed microfluidic devices: Enablers and barriers", *Lab Chip* 2016, 16, 1993. **DOI: 10.1039/c6lc00284f.**

[6]   M. Trebbin, K. Krüger, D. DePonte, S. V. Roth, H. N. Chapman, S. Förster, "Microfluidic liquid jet system with compatibility for atmospheric and high-vacuum conditions", *Lab Chip* 2014, *14*, 1733. **DOI: 10.1039/c3lc51363g**.

[7]   L. Jiang, Y. Zeng, Q. Sun, Y. Sun, Z. Guo, J. Y. Qu, S. Yao, "Microsecond protein folding events revealed by time-resolved fluorescence resonance energy transfer in a microfluidic mixer", *Anal. Chem.* 2015, *87*, 5589. **DOI: 10.1021/acs.analchem.5b00366**.

[8]   D. J. Kim, H. J. Oh, T. H. Park, J. B. Choo, S. H. Lee, "An easily integrative and efficient micromixer and its application to the spectroscopic detection of glucose-catalyst reactions", *Analyst* 2005, *130*, 293. **DOI: 10.1039/b414180f**.

[9]   Y. Gambin, C. Simonnet, V. van Delinder, A. Deniz, A. Groisman, "Ultrafast microfluidic mixer with three-dimensional flow focusing for studies of biochemical kinetics", *Lab Chip* 2010, *10*, 598. **DOI: 10.1039/B914174J**.

[10]   J. B. Knight, A. Vishwanath, J. P. Brody, R. H. Austin, "Hydrodynamic focusing on a silicon chip: Mixing nanoliters in microseconds", *Phys. Rev. Lett.* 1998, *80*, 3863. **DOI: 10.1103/PhysRevLett.80.3863**.

[11]   A. A. S. Bhagat, E. T. K. Peterson, I. Papaustky, "A passive planar micromixer with obstructions for mixing at low Reynolds numbers", *J. Microchem. Microeng.* 2007, *17*, 1017. **DOI: 10.1088/0960-1317/17/5/023**.

[12]   J. Cha, J. Kim, S.-K. Ryu, J. Park, Y. Jeong, S. Park, S. Park, H. C. Kim, K. Chun, "A highly efficient 3D micromixer using soft PDMS bonding", *J. Microchem. Microeng.* 2006, *16*, 1778. **DOI: 10.1088/0960-1317/16/9/004**.

[13]   N.-T. Nguyen, Z. Wu, "Micromixers – a review", *J. Microchem. Microeng.* 2005, *15*, R1. **DOI: 10.1088/0960-1317/15/2/R01**.

[14]   N. Aoki, S. Hasebe, K. Mae, "Mixing in microreactors: effectiveness of lamination segments as a form of feed on product distribution for multiple reactions", *Chem. Eng. J.* 2004, *101*, 323. **DOI: 10.1016/j.cej.2003.10.015**.

[15]   M. Trebbin, D. Steinhauser, J. Perlich, A. Buffet, S. V. Roth, W. Zimmermann, J. Thiele, S. Förster, "Anisotropic particles align perpendicular to the flow direction in narrow microchannels", *Proc. Natl. Acad. Sci USA* 2013, *110*, 6706. **DOI: 10.1073/pnas.1219340110**.

[16]   A. Rotem, A. R. Abate, A. S. Utada, V. van Steijn, D. A. Weitz, "Drop formation in non-planar microfluidic devices", *Lab Chip* **2012**, *12*, 4263. **DOI: 10.1039/c2lc40546f**.

[17]   S. Devasenathipathy, J. G. Santiago, "Electrokinetic flow diagnostics", in *Microscale Diagnostic Techniques* **2005**, ed. K. S. Breuer, Springer Heidelberg, 113. **DOI: 10.1007/b137604**.

[18]   J. I. Molho, "Electrokinetic Dispersion in Microfluidic Separation Systems", PhD. Thesis, Stanford University, 2001.

[19]   D. Juncker, H. Schmid, U. Drechsler, H. Wolf, M. Wolf, B. Michel, N. de Rooij, E. Delamarche, "Autonomous microfluidic capillary system", *Anal. Chem.* 2002, *74*, 6139. **DOI: 10.1021/ac0261449**.

[20]   M. Zimmermann, H. Schmid, P. Hunziker, E. Delamarche, "Capillary pumps for autonomous capillary systems", *Lab Chip* 2007, *7*, 119. **DOI: 10.1039/b609813d**.

[21]   A. R. Abate, D. A. Weitz, "Syringe-vacuum microfluidics: A portable technique to create monodisperse emulsions", *Biomicrofluidics* 2011, *5*, 014107. **DOI: 10.1063/1.3567093**.

[22]   J. S. Pöll, "The story of the gauge", *Anaesthesia* 1999, *54*, 575. **DOI: 10.1046/j.1365-2044.1999.00895.x**.

[23]   M. Hoffmann, M. Schlüter, N. Räbiger, "Experimental investigation of liquid-liquid mixing in T-shaped micromixers using μ-LIF and μ-PIV", *Chem. Eng. Sci.* 2006, *61*, 2968. **DOI: 10.1016/j.ces.2005.11.029**.

[24]   S. Seiffert, "Origin of nanostructural inhomogeneity in polymer-network gels", *Polymer Chem.* 2017, *8*, 4472. **DOI: 10.1039/C7PY01035D**.

[25] A. Groisman, M. Enzelberger, S. R. Quake, "Microfluidic memory and control devices", *Science* 2003, *300*, 955. **DOI: 10.1126/science.1083694**.
[26] A. Gang, N. Haustein, L. Baraban, W. Weber, T. Mikolajick, J. Thiele, G. Cuniberti, "Microfluidic alignment and trapping of 1D nanostructures – a simple fabrication route for single-nanowire field effect transistors", *RSC Advances* 2015, *5*, 94702. **DOI: 10.1039/C5RA20414C**.
[27] M. Antonietti, S. Förster, "Vesicles and liposomes: A self-assembly principle beyond lipids", *Adv. Mater.* 2003, *15*, 1323. **DOI: 10.1002/adma.200300010**.
[28] A. Benedetto, G. Accetta, Y. Fujita, "Spatiotemporal control of gene expression using microfluidics", *Lab Chip* 2014, *14*, 1336. **DOI: 10.1039/c3lc51281a**.
[29] A. R. Abate, D. Lee, T. Do, C. Holtze, D. A. Weitz, "Glass coating for PDMS microfluidic channels by sol-gel methods", *Lab Chip* 2008, *8*, 516. **DOI: 10.1039/b800001h**.
[30] A. R. Abate, A. T. Krummel, D. Lee, M. Marquez, C. Holtze, D. A. Weitz, "Photoreactive coating for high-contrast spatial patterning of microfluidic device wettability", *Lab Chip* 2008, *8*, 2157. **DOI**: 10.1039/b813405g.
[31] A. R. Abate, J. Thiele, M. Weinhart, D. A. Weitz, "Patterning microfluidic device wettability using flow confinement", *Lab Chip* 2010, *10*, 1774. **DOI: 10.1039/c004124f**.
[32] W.-A. C. Bauer, M. Fischlechner, C. Abell, W. T. S. Huck, "Hydrophilic PDMS microchannels for high-throughput formation of oil-in-water microdroplets and water-in-oil-in-water double emulsions", *Lab Chip* 2010, *10*, 184. **DOI: 10.1039/c004046k**.
[33] S. C. Kim, D. J. Sukovich, A. R. Abate, "Patterning microfluidic device wettability with spatially-controlled plasma oxidation", *Lab Chip* 2015, *15*, 3163. **DOI**: 10.1039/c5lc00626k.
[34] S. Seiffert, F. Friess, A. Lendlein, C. Wischke, "Faster droplet production by delayed surfactant-addition in two-phase microfluidics to form thermo-sensitive microgels", *J. Colloid Interf. Sci.* 2015, *452*, 38. **DOI: 10.1016/j.jcis.2015.04.017**.
[35] D. J. Holt, R. J. Payne, C. Abell, "Synthesis of novel fluorous surfactants for microdroplet stabilisation in fluorous oil streams", *J. Fluorine Chem.* 2010, *131*, 398. **DOI: 10.1016/j.jfluchem.2009.12.010**.
[36] D. J. Holt, R. J. Payne, W. Y. Chow, C. Abell, "Fluorosurfactants for microdroplets: Interfacial tension analysis", *J. Colloid Interface Sci.* **2010**, *350*, 205. **DOI: 10.1016/j.jcis.2010.06.036**.
[37] L. S. Roach, H. Song, R. F. Ismalgilov, "Controlling nonspecific protein adsorption in a plug-based microfluidic system by controlling interfacial chemistry using fluorous-phase surfactants", *Anal. Chem.* 2005, *77*, 785. **DOI: 10.1021/ac049061w**.
[38] A. R. Abate, D. A. Weitz, "Faster multiple emulsification with drop splitting", *Lab Chip* 2011, *11*, 1911. **DOI: 10.1039/c0lc00706d**.
[39] Y. Ma, J. Thiele, L. Abdelmohsen, J. Xu, W. T. S. Huck, "Biocompatible macro-initiators controlling radical retention in microfluidic on-chip photo-polymerization of water-in-oil emulsions", *Chem. Commun.* 2014, *50*, 112. **DOI: 10.1039/c3cc46733c**.
[40] S. Seiffert, D. A. Weitz, "Controlled fabrication of polymer microgels by polymer-analogous gelation in droplet microfluidics", *Soft Matter* 2010, *6*, 3184. **DOI: 10.1039/C0SM00071J**.
[41] J. Thiele, S. Seiffert, "Double emulsions with controlled morphology by microgel scaffolding", *Lab Chip* 2011, *11*, 3188. **DOI: 10.1039/C1LC20242A**.
[42] N. Bremond, H. Doméjean, J. Bibette, "Propagation of drop coalescence in a two-dimensional emulsion: A route towards phase inversion", *Phys. Rev. Lett.* **2011**, *106*, 214502. **DOI: 10.1103/PhysRevLett.106.214502**.

Appendix

We would like to close this book with expressing our sincere gratitude to colleagues, friends, and our families who provided support, a store of information, images and reports from their own work, and who served as critical proofreaders of our manuscript during the time of its evolution. In particular, we thank...

- Julia Lauterbach, Ria Fritz, and Lena Stoll (DeGruyter) for inspiring us to write this book, for technical guidance, and for supervision of the text book's editing process
- Dr. Carola Graf, Nicolas Hauck, Max J. Männel, and Jens W. Neubauer (Leibniz Institute of Polymer Research Dresden) for providing data and microscopy images of microfluidic experiments used in Chapter 3
- Camilla Scheibe (Leibniz Institute of Polymer Research Dresden) for designing sketches of 3D microchannels and velocity profiles
- Dr. Martin Trebbin (University of Buffalo) for providing FEM simulations used in Chapter 1
- Emanuel Richter (Leibniz Institute of Polymer Research Dresden) and David Rochette (JGU Mainz) for assisting with photography of some microfluidic setups and procedures shown in this book
- Participants of the master course "Introduction to Bionanotechnology" in the summer terms of 2017 and 2018 (BIOTEC, TU Dresden) for their critical feedback on didactical aspects of this book
- Our wives Sara and Franziska for their endless patience with us in the vibrant time of writing this book in the years 2016–2019, at the side of our main duties in research and teaching at JGU Mainz, IPF Dresden, and TU Dresden, and at the side of the birth of our children Levke (2016), Lea (2017), and Arved (2018).

Index

active mixer 155
additive manufacturing 101
Aquapel 170, 254
aspect ratio 150
AutoCAD 144, 251

backflow 152, 198
baker's transformation 82
biopsy 166
Buckingham pi theorem 11

capillary device 67
capillary forces 187
capillary number 64, 69
capillary pump 189
capillary stress 66
capsule 92
centipoise 6
channel junction 149
chaotic advection 82, 155
circular glass capillary 174
clean room 141
clogging 141, 151
coalescence 59
coating 99
coaxial alignment 174
co-flow 63
co-flow geometry 96, 175
colloidosome 108
colloids 26
colloidsl dispersion 34
complex fluid 26
computational fluid dynamics simulation 15
computer-aided design (CAD) 142
contact angle 56
continuity equation 4
continuous phase 78
core-shell capsule 86
Couette flow 5
critical micelle concentration 61
cross-junction channel 87, 97

dead volume 180, 191
dead zone 151
delay line 149
developer bath 162

dewetting transition 113
diffusion 37
digital camera 195
dimensionless number 11
dispersed phase 78
dissolution 178
double emulsion 86
dripping 66, 96
dripping regime 66
dripping-to-jetting transition 68, 78
drop maker 86
droplet-based microfluidics 77
dust 141

Einstein equation 34, 42
Einstein-Smoluchowski equation 31, 47
elastomer 97, 142
electro-osmotic flow 186
emulsion 59
encapsulation 85, 106
epoxy glue 174

faucet 65
Fick's first law 20, 38
Fick's second law 38, 43
filter unit 149
finite element method 15
flow confinement 221
flow field 2
flow-focusing geometry 67, 97, 175
flow-focusing junction 62, 78, 87
flow rate 14, 65, 93
flow-rate ratio 154, 199, 236
fluid inlet 149
fluid outlet 149
fluorescence microscope 193
friction 5
frictional force 1

gauge 191
geometric mixer 155
geometry-controlled droplet breakup 128
glass-capillary device 86
glass-capillary microfluidics 14
grafting polymerization 217

Hagen-Poiseulle law 7
hard bake 162
HFE 7500 231
H-filter 157
high-speed camera 195
high-throughput screeing 103
hydrodynamic flow-focusing 87, 154, 208
hydrodynamic radius 30, 42
hydrodynamic resistance 130
hydrophilic 96
hydrophobic 95

ideal fluid 1
in-field analytics 205
inner fluid 86
integrated device 205
interface 52
interfacial tension 53, 56, 65, 66
inverse microscope 193

Janus 83, 106, 241
jetting 67

Kapton 141, 172
Krytox 235

lab on a chip 205
laminar flow 2, 11
Laplace pressure 55
layer-by-layer 99, 227
liposome 108

Marangoni convection 70
mask aligner 140, 147, 160
meander 149
mean-square displacement 47
microcapillaries 66
microcapsule 108, 245
microchannel network 144
microgel 104, 239
micromixer 155
micropipette puling machine 177
middle fluid 86
mixing 1, 154
multiple emulsion 86

nanoprecipitation 210
Navier-Stokes equation 10
Newtoian fluid 27

Newtonian flow 27
nonspherical 106, 240
Nowton's law 6
nucleation 150
numbering-up 123, 173

one-step 90, 245
orifice 66, 96
Ostwald ripening 55
outer fluid 86
overlap concentration 30, 32
oxygen plasma 127

packing parameter 213
parabolic flow profile 150, 186, 205
paraffin oil 231
parallelization 123, 173
partial differential equations 14
passive mixer 155
PDMS 87, 142, 164
Péclet number 25
photolithography 97, 158
photomask 142, 144, 146, 161
photopatterning 172, 221
photoresist 158
pipette puller 174
plasma treatment 127, 167
Poiseuille flow 5, 12
Poisson distribution 246
poly(dimethylsiloxance) 97, 142, 164
polymer solution 29
polymers 26
polymersome 109, 211
post-exposure bake 161
pressure-driven flow 150, 184
punch hole 166

random walk 45
rapid prototyping 15, 142
Rayleigh number 49
Rayleigh-Plateau instability 68
relaxation time 26
replica 142, 172
residence time 150
Reynolds number 11

Schmidt number 26
segmented flow 63, 77
self-assembly 150

sensor 205
sequencing 103
shear thickening 36
shear thining 28
shell thickness 91
silicone elastomer 142
simulation 142
single emulsion 84
soda lime 147
soft bake 159
soft lithography 15, 18, 120, 164
soft matter 26
sol-gel coating 217, 219, 221
sol-gel layer 217
spreading parameter 113
square glass capillary 174
stamped microfluidics 14
stationary 1
stationary concentration profile 154
Stokes-Einstein equation 42
Stokes equation 12
streamline 2
strobing 196
SU-8 158
surface polymerization 99
surface tension 52
surfactant 58
Sylgard elastomer construction kit 164

syringe pump 184
syringe tip 174

tapered 96
Taylor dispersion 26
template 104
throughput 123
T-junction 66
T-shaped 149
turbulent flow 2

upright microscope 193

vesicle 109
viscoelasticity 26
viscosity 6
viscous fluid 1
viscous stress 66

Weber number 69
Weissenberg number 27
wettability 90, 97, 170, 216
wettability control 99
wettability pattern 127, 170, 223
wettability treatment 142, 174
wetting 1, 55, 78
wetting defect 201

X-shaped 149

Y-shaped 149